陳雲潮 編著

iLAB Digital
數位電路設計、模擬測試與硬體除錯

東華書局

國家圖書館出版品預行編目資料

iLAB Digital 數位電路設計、模擬測試與硬體除錯 / 陳雲潮編著. -- 1 版. -- 臺北市 : 臺灣東華, 2015.11

264 面 ; 17x23 公分.

ISBN 978-957-483-848-6 (平裝)

1. 積體電路 2. 設計

448.62　　　　　　　　　　　　　104024300

iLAB Digital 數位電路設計、模擬測試與硬體除錯

編 著 者	陳雲潮
發 行 人	陳錦煌
出 版 者	臺灣東華書局股份有限公司
地　　址	臺北市重慶南路一段一四七號三樓
電　　話	(02) 2311-4027
傳　　真	(02) 2311-6615
劃撥帳號	00064813
網　　址	www.tunghua.com.tw
讀者服務	service@tunghua.com.tw
門　　市	臺北市重慶南路一段一四七號一樓
電　　話	(02) 2371-9320
出版日期	2015 年 11 月 1 版 1 刷
	2020 年 2 月 1 版 2 刷

ISBN　　978-957-483-848-6

版權所有 ‧ 翻印必究

推薦序

校友陳雲潮先生，早年畢業於臺北工專電機科，並曾擔任電子科助教、講師、副教授兼電子科主任等職多年，1972 年赴美後，於 General Instruments、Texas Instruments 及 IBM 等公司擔任資深工程師共三十餘年。退休後受聘於金門大學擔任講座教授五年，回臺北後復於母校電子工程系兼課。

陳老師深感學生在校實作時數不足，以致畢業後無法配合工業界之需求，2012 年自美國引進 Digilent 公司新推出的 Analog Discovery Module，價格較低、學生足以負擔，也可以讓學生把電子實驗帶回家做。我回憶起以前當學生的時候因為示波器過於昂貴，只有在實驗室才有機會學習示波器的操作，所以在電子電路量測及設計實作技能上的學習有限。如果學生能擁有這樣一個可以負擔得起的設備，相信必能在家裡配合自己的電腦，不受空間及在校實作時數不足的限制，大幅強化電子電路量測及設計的實作技能。因此，我拜託陳教授將這個綜合電子測試裝置的實作學習納入本校電子工程系實驗課程。課程一開始由陳老師教授，陳老師非常愛護本校教師與學生，關心學生實作技能的培養，並且樂意將教學內容傳承給其他老師，使本課程得以繼續開設。

為了配合這個 Module，陳老師非常用心，重新編寫實驗講義，將 C/C++ Programing、模擬測試、硬體組裝和除錯集合在一起，稱之為 iLAB。先後完成了 iLAB-Analog 和 iLAB-Digital 二書，由東華書局出版試用版，再經試教後始完成修正本。

陳老師以 82 歲之高齡仍不忘將畢生所學薪火相傳，如今本書正式出版，將使此一領域的學生在實作技術的起跑點上，獲得一大躍進，謹以此序向其表達感謝之意。

國立臺北科技大學校長

姚立德　　敬書

2015 年 2 月 3 日

推薦序

我任職於一元素科技公司多年，負責 Xilinx 大學計劃工作，期間見過不少優秀的老師，但陳雲潮老師卻是我見過印象最深刻的一位。2013 年四月初，我收到原廠德致倫先生的 Email 通知，請代理商盡速聯繫並協助一位來自台灣教授所提出的需求，這封郵件就是來自我母校臺北科技大學。民國 87 年，我畢業於臺北科技大學電子系，工作幾年後，再度回到母校修讀碩士，由於多年與學校的感情，自然非常關心這封郵件。

記得第一次和陳老師見面，一位個子瘦瘦小小、年歲非常高的教授，面帶笑容十分的親和。經過幾次見面並討論，對老師有更進一步的認識。

老師於 2000 年自美國紐約州 IBM 公司退休後，於 2005 年回臺灣。先在金門技術學院擔任了五年的講座教授。三年前回母校電子系兼任教授。在旅美數十年的工程師生涯中，從未中斷在大學裡兼任授課。隨時在準備自己，總憧憬著有朝一日能回到自己的國家，為後輩學子盡份棉薄之力。二年前暑假去美國，注意到美國大學開始採用 Analog Discovery，現在在校園裡已非常普遍，而且很是受到歡迎，有關科系的學生幾乎人手一機的地步。有鑑於此，老師自己買了一個在家裡實地操作，發覺它非常實用，假如能普及和推廣，

v

簡直就是革命性的改變以往實習課程僅限於固定的時數和實驗室空間的傳統觀念，只要把 Analog Discovery 帶回家就能操作，甚至在咖啡廳也能輕鬆完成。於是他決定寫一本能輔助實作的書。為了趕寫這本書，課本中所有的實驗，都是老師親自設計，並且走到光華商場買零件，一個一個實做完成。為了與時間賽跑，真可說是日以繼夜，一頁頁的往下寫。先印試教本，等一學期教授完畢，依據同學吸收的結果，再修改成永久版。老師為了讓同學們能輕鬆的購買到書本，以台幣一元的金額，將書的版權賣給出版商，可謂是用心良苦啊！

在校長大力的支持之下，老師找到了一元素，希望能給予學校和學生最優惠實驗工具的價格，共同來幫助老師圓夢。所以在這本書的背後，隱藏著無數的付出、努力和用心，各位讀者以及正在使用這本書的學生們，希望這本書除了增長你們的知識外，也讓你們瞭解到何謂真正的"教育家"。

黃裕鈞

前言

　　iLAB-Digital 與 iLAB-Analog 類似，爲使用 Analog Discovery 的實作講義。專門爲北科大電子系設計的初等邏輯實作教材。全書共 13 章，供一個學期，每週 3 小時之用。其中第一章到第六章爲基本邏輯電路，包含 Logic Gates、Adder、JKFF、MUX/DEMUX、Shift Register 和 Counter。第七章到第十一章，包含 Digital to Analog Conversion、Analog to Digital Conversion、Clock and PLL、Stepper and Driver 和 Servor and Control，第十二和十三章爲 Altera 和 Xilinx FPGA 對基本邏輯電路的合成。

　　每章的結構爲：

1. 電路構成的描述。
2. VHDL 對電路構成的描述與模擬測試。
3. 電路結構的佈置與實作。
4. 使用 Analog Discovery 對電路的測試與除錯。
5. 課外練習。

時間的分配上，其中：

　　(1) 由教師講述講義中 1~3 項的範例及解答疑問，大約佔用 1 小時。

　　(2) 教師講述之後，同學需重複練習第 2 項中的範例，最多也佔用 1

小時。

(3) 同學須重複練習第 3~4 項中的範例，最多也佔用 1 小時。

(4) 同學除了須在課外完成第 4 項之課外練習外，同時須完成在課內未完成之模擬測試與 Analog Discovery 對電路的測試與除錯等項目，大約為 4~8 小時。

實作使用之工具與器材：

1. 軟體部份：VHDL Simulator 模擬測試用軟體，可從 Altera/Quartus 網站下載。Synthesis 部份，如果使用的是 Xilinx FPGA，可從 Xilinx/ISE 網站下載。Waveform 測試用軟體，可從 Digilent 的網站下載。
2. 硬體部份：第一章到第十一章的電子零件，都是普通的低價位零件，全省各大都市的電子商場都有提供。第十二章的 FPGA 實作平台為 Altera/Terasic 的 DE2-115。第十三章的 FPGA 實作平台為 Xilinx/Digilent 的 BASYS3。由於 iLAB-Digital 使用 Digilent/Analog Discovery 的 Digital Pattern Generator 和 Logic Analyzer 來做測試，幾乎不受 FPGA 實作平台的影響。

北科大電子系

陳雲潮

目次

推薦序	....	iii
推薦序	....	v
前言	....	vii
目次	....	ix

第一章　使用 NAND 邏輯閘來合成其它 Logic Gates

1-1	使用 2-input NAND 來構成其它 Logic Gates 的電路	1
1-2	使用 VHDL 來描述電路的結構和電路的模擬測試	2
1-3	硬體實作	4
1-4	課外練習	8

第二章　加法器電路的組成和測試

2-1	簡單的加法器電路	9
2-2	VHDL Coding 與 ModelSim Simulation	10
2-3	Carry-ripple adder SN74LS283	12
2-4	Carry-Lookahead adder	13
2-5	硬體實作	14
2-6	課外練習	19

第三章　JKFF 電路的組成和測試

3-1	JKFF 電路的組成	21
3-2	VHDL Coding 與 ModelSim Simulation	23

ix

	3-3	硬體實作	25
	3-4	課外練習	29
	3-5	註釋	29

第四章　Data Selectors 與 Multiplexers / Demultiplexers

	4-1	4-line to 1-line Data Selectors/Multiplexers	31
	4-2	3-line to 8-line Decoder/Demultiplexers	33
	4-3	MUX 和 DECODER 的測試	34
	4-4	MUX 電路在 FPGA 中的用途	37
	4-5	硬體實作	38
	4-6	課外練習	43

第五章　移位寄存器

	5-1	DFF 組成的移位寄存器	45
	5-2	使用 VHDL 來描述電路的結構和電路的模擬測試	46
	5-3	雙向 Shift Register SN74LS194 的介紹	48
	5-4	硬體實作	50
	5-5	課外練習	54

第六章　計數器

	6-1	DFF 組成的計數器	55
	6-2	使用 VHDL 來描述電路的結構和電路的模擬測試	56
	6-3	74LS193 Synchronous 4-Bit Binary Counter with Dual Clock 簡介	58
	6-4	硬體實作	60
	6-5	課外練習	65

目次　xi

第七章　數位訊號轉換成類比訊號

- 7-1　階梯式 R2R 電阻組成的 D to A 轉換器 67
- 7-2　D to A 積體電路 DAC0808 的介紹 71
- 7-3　硬體實作 73
- 7-4　課外練習 80

第八章　類比訊號轉換成數位訊號

- 8-1　ADC 0804 的介紹 81
- 8-2　ADC0804 的測試 84
- 8-3　硬體實作 85
- 8-4　課外練習 91

第九章　Clock Generation 與 PLL

- 9-1　簡單時序脈波的產生 93
- 9-2　Clock 的分佈 93
- 9-3　不同頻率的要求 94
- 9-4　鎖相環 PLL 電路的結構 94
- 9-5　相位檢測器 95
- 9-6　PLL 的 Clock 頻率倍增器 98
- 9-7　LM565/PLL 和 74LS90/Decade Counter 構成的 x10 倍頻器 98
- 9-8　硬體實作 99
- 9-9　課外練習 103

第十章　Step Motor 與 Driver

- 10-1　單相 Unipolar 步進馬達 105
- 10-2　Bipolar 步進馬達 107
- 10-3　5804 IC 以推動 4 組線圈的步進馬達 107

- 10-4 微小型 Unipolar 步進馬達的規格 108
- 10-5 硬體實作 109
- 10-6 課外練習 116

第十一章　Servo System 與 Control

- 11-1 伺服系統的內部結構 117
- 11-2 伺服系統的控制 118
- 11-3 Micro Servo MG90S 的規格與連線 119
- 11-4 Servo Control IC 121
- 11-5 硬體實作 122
- 11-6 課外練習 124

第十二章　Altera/Quartus DE2-115 FPGA 的電路合成

- 12-1 DE2-115 的 Expansion Header 126
- 12-2 Synthesis 軟體 Quartus II / VHDL 的介紹 129
- 12-3 Synthesis 電路的一個例子 131
- 12-4 Synthesis FPGA 的介紹 151
- 12-5 Analog Discovery 對 DE2-115 FPGA 的測試 ... 161
- 12-6 課外練習 164

第十三章　Xilinx/Vivado Basys3 FPGA 的電路合成

- 13-1 Vivado 軟體的運作介紹 165
- 13-2 Vivado 的 Simulator 運作介紹 177
- 13-3 Vivado 軟體的 Implementation 和 Bitstream 檔案的產生 185
- 13-4 Vivado 軟體的 Bitstream Programming Basys3 191
- 13-5 Analog Discovery 對 Basys3/FPGA Counter 的測試 ... 193
- 13-6 課外練習 195

附錄

- 附錄 A　VHDL 電路檔的結構與格式 197
- 附錄 B　VHDL 測試檔的結構與 Stimulus 的寫法 200
- 附錄 C　如何使用和產生 Digital Pattern 204
- 附錄 D　如何使用 Logic Analyzer 212
- 附錄 E　ModelSim Simulation 模擬測試 215
- 附錄 F　LM565 PLL 的設定 225
- 附錄 G　Xilinx CPLD 的電路合成 229

索引

中英對照 247
英中對照 248

第一章　使用 NAND 邏輯閘來合成其它 Logic Gates

邏輯閘是構成一切數位電路的基本元件。邏輯閘被稱為**並發邏輯** (Concurrent Logic)，它不同於**時序邏輯** (Sequential Logic)，依靠系統中的**時鐘** (Clock) 來行事。邏輯閘的種類很多，但卻互通。只須用任何一種含有 NOT 的邏輯閘，便可以構成各種其它邏輯。

1-1　使用 2-input NAND 來構成其它 Logic Gates 的電路

圖 1-1 是使用 2-input NAND 來構成其它 Logic Gates 的電路。

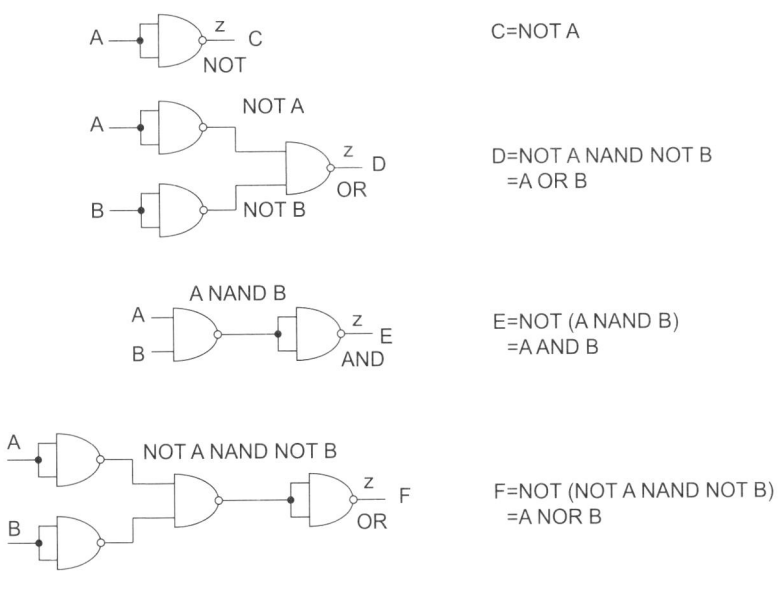

圖 1-1：　使用 2-input NAND 來構成其它 Logic Gates 的電路

1

1-2 使用 VHDL 來描述電路的結構和電路的模擬測試

1-2-1 圖 1-2 為圖 1-1 的 VHDL 電路的結構描述 (請參考附錄 A)

```
1  LIBRARY IEEE;
2  USE IEEE.STD_LOGIC_1164.ALL;
3  ENTITY SN74LS00 IS
4      PORT( A, B: IN STD_LOGIC;
5             C_NOT, D_OR, E_AND, F_NOR: OUT STD_LOGIC);
6  END ENTITY;
7  ARCHITECTURE structure OF SN74LS00 IS
8  BEGIN
9      U1: C_NOT <= NOT A;
10     U2: D_OR  <= NOT A NAND NOT B;
11     U3: E_AND <= NOT (A NAND B);
12     U4: F_NOR <= NOT(NOT A NAND NOT B);
13 END structure;
```

圖 1-2　　VHDL 電路的結構描述

1-2-2 圖 1-3 為圖 1-2 的 VHDL 測試檔 (請參考附錄 B)

```
1  -- testbench for SN74LS00
2  Library IEEE;
3  USE IEEE.STD_LOGIC_1164.ALL;
4
5  ENTITY testbench IS
6  END testbench;
7
8  USE work.all;
9
10 ARCHITECTURE stimulus of testbench IS
11     component SN74LS00
12     PORT( A, B: IN STD_LOGIC;
13            C_NOT, D_OR, E_AND, F_NOR: OUT STD_LOGIC);
14     end component;
15 -- Declare testbench's signals
16 SIGNAL A, B: STD_LOGIC :='0';
17 SIGNAL C_NOT, D_OR, E_AND, F_NOR: STD_LOGIC;
18
19 BEGIN
20     DUT: SN74LS00 PORT MAP(A, B, C_NOT, D_OR, E_AND, F_NOR);
21
22 --  test signal generation:
23     A <= NOT A AFTER 100 ns;
24     B <= NOT B AFTER 200 ns;
25 END stimulus;
```

圖 1-3　　圖 1-2 的 VHDL 測試檔

如果使用 ModelSim Simulator，把圖 1-2 和圖 1-3 來做模擬測試，可以獲得如圖 1-4 的信號的時序圖，來驗證設計的結果。 (請參考附錄 E)

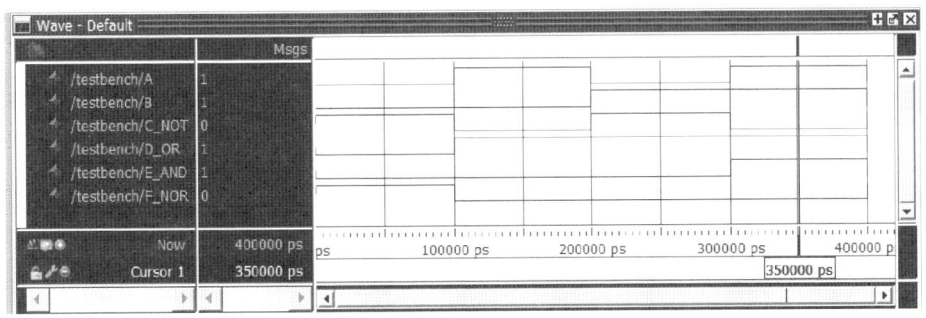

　　圖 1-4　　　　電路信號的時序圖，驗證設計的結果

波形的解讀：如圖 1-4 所示，以 1,000,000 ps (1 ms) 作為時間的單位，則 A, B, C, D, E, F 等 6 個訊號以 Binary 橫向來表示當為：

A : 0, 1, 0, 1 → A
B : 0, 0, 1, 1 → B
C : 1, 0, 1, 0 → NOT A
D : 0, 1, 1, 1 → A OR B
E : 0, 0, 0, 1 → A AND B
F : 1, 0, 0, 0 → A NOR B

1-3 硬體實作

TTL 的 74LS00 是一只含有 4 個 2-input 的 NANDs，使用 2 只 74LS00 中的 6 個 2-input 的 NANDs，就可以完成圖 1-1 的工作，如圖 1-5 所示。

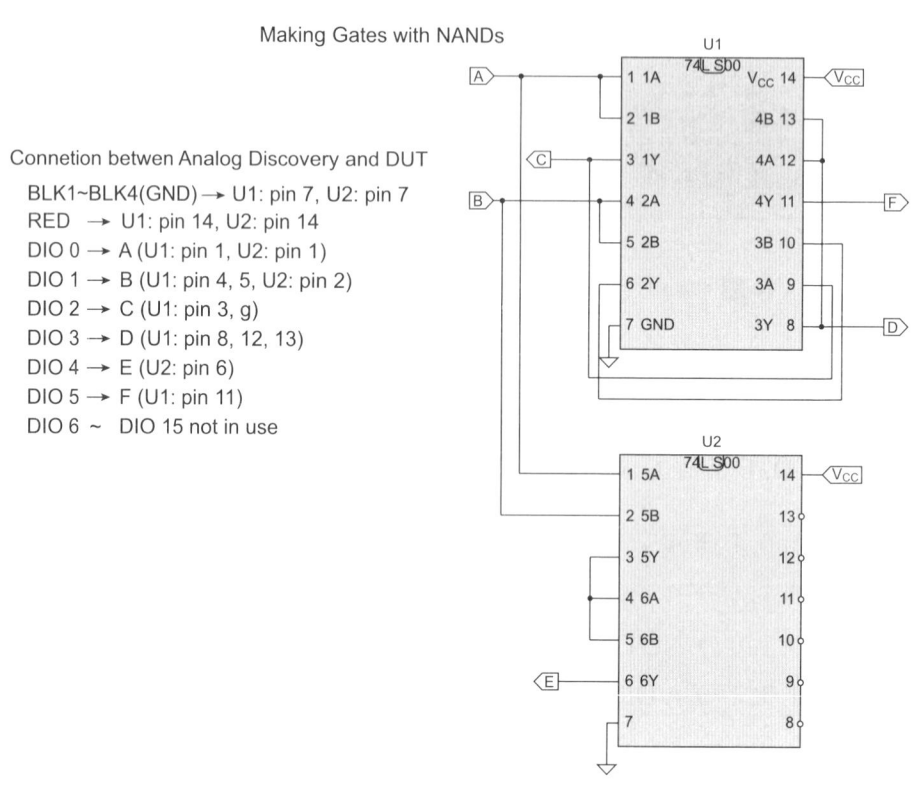

圖 1-5　74LS00 NANDs 組成其它 Gates

簡單的數位電路，如果使用比較低的頻率 (1 KHz)，可以使用不用銲接的**麵包板** (Breadboard) 來完成，如圖 1-6 所示。頻率較高 (10 KHz) 以上或比較複雜的電路，應當把電路裝配在接地良好、使用銲接的電路板上。

第一章 使用 NAND 邏輯閘來合成其它 Logic Gates 5

圖 1-6　將電路配置在不用銲接的麵包板上

　　圖 1-6 麵包板上的電路，經由接線柱，連接到 Analog Discovery Module，連同裝上了 WaveForms 軟體的 PC，成為多件由軟體操控的硬體儀器。測試數位電路，對於電路的輸入部份可用**數字模式產生器** (Digital Pattern Generator)，輸出部份可用**數位信號分析儀** (Digital Signal Analyzer)，圖 1-7 就是 WaveForms 軟體在 PC 上所顯示的選擇圖形。

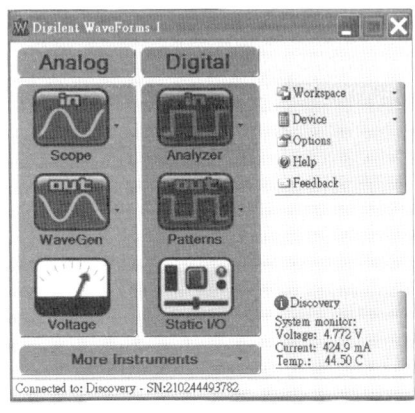

圖 1-7　WaveForms 軟體在 PC 上所顯示的選擇圖形

點擊 Digital 部份的 out，便可獲得如圖 1-8 的數字模式產生器。如何使用 Digital Pattern Generator 和產生 Digital Pattern？(請參考附錄 C)

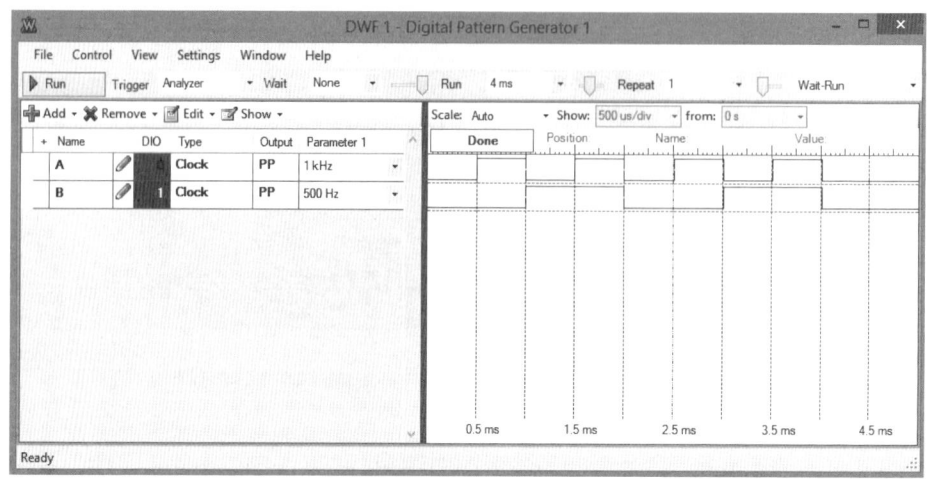

圖 1-8　　提供電路輸入訊號的數字模式產生器

點擊 Digital 部份的 in，便可獲得如圖 1-9 的數位信號分析儀。如何使用 Logic Analyzer？(請參考附錄 D)

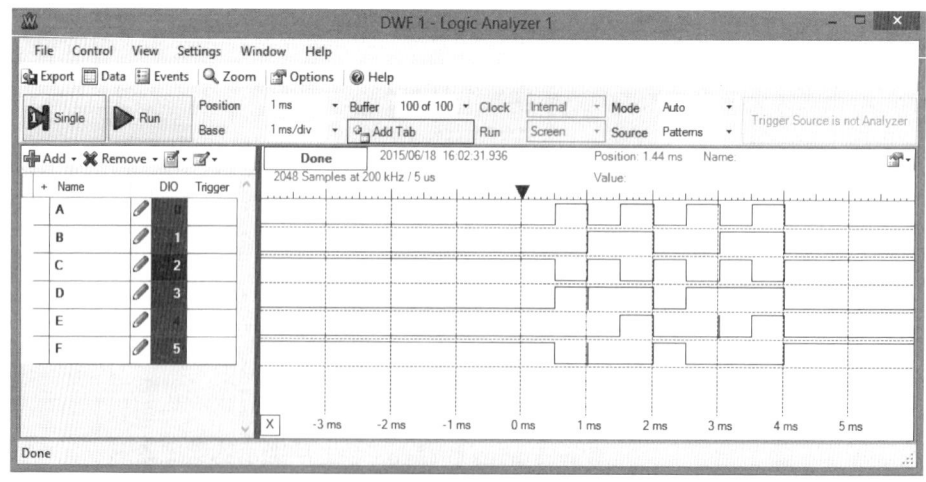

圖 1-9　　提供電路記錄輸入訊號的數位信號分析儀

最後，點擊 **Analog** 部份的 **Voltage**，以便 SN74LS00 可獲得如圖 1-10 的 2 個固定 ±5 V 電壓源。TTL IC 使用 V+ 即可。V- OFF。

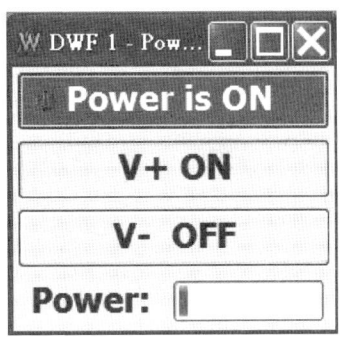

圖 1-10　　二個固定 ±5 V 的電壓源

1-4 課外練習

1. 試將 2-input NOR Gates 來組成 NOT, AND, NAND, OR 等 Gates。
   A. 寫出電路的 VHDL 結構編碼。
   B. 用 ModelSim Simulator 來測試證實。
   C. 用 CMOS IC 4001 實作並使用 Analog Discovery 來測試證實。

2. 下圖為 Odd Parity Detector 的電路，試寫出它的 VHDL 模式及 TestBench 並用 ModelSim Simulator 測試之。

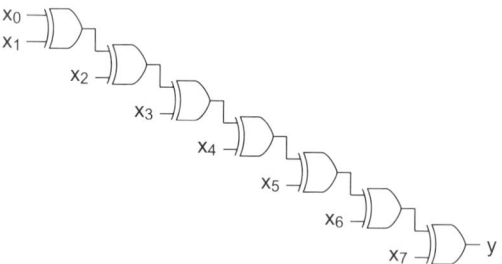

Odd Parity Detector 電路

3. 下圖為 2 to 1 Multiplex 的電路，試寫出它的 VHDL 模式及 TestBench 並用 ModelSim Simulator 測試之。

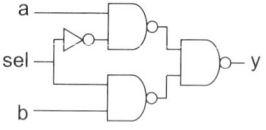

2 to 1 Multiplex 電路

4. 下圖為將 2 to 1 Multiplex 加上回授而成的電路，試問它還是 Combinational Logic 嗎？

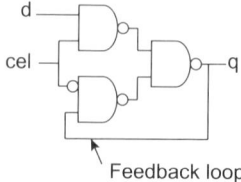

2 to 1 Multiplex 加上回授而成的電路

第二章 加法器電路的組成和測試

加法器電路是決定一部計算機速度的基本元件。組成這個元件的電路有很多種，最簡單同時也最慢的是 Carry-ripple adder，比較快也比較複雜的是 Carry-Lookahead adder。本章介紹它們的組成、VHDL Coding、Simulation 和硬體測試的方法。

2-1 簡單的加法器電路

圖 2-1 是加法器的方塊圖和它的 Truth Table。所有的加法器電路設計皆是從這裡開始。

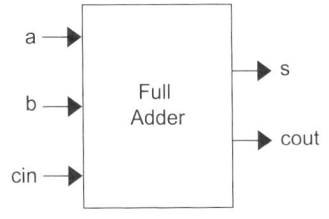

a	b	cin	s	cout
0	0	0	0	0
0	1	0	1	0
1	0	0	1	0
1	1	0	0	1
0	0	1	1	0
0	1	1	0	1
1	0	1	0	1
1	1	1	1	1

圖 2-1： 加法器的方塊圖和它的 Truth Table

從 Truth Table 可以寫出加法器的輸出 Sum (S) 和 Carry OUT (cout)，跟輸入 a、b 和 Carry IN (cin) 的 Logic 關係。如圖 2-2 所示。

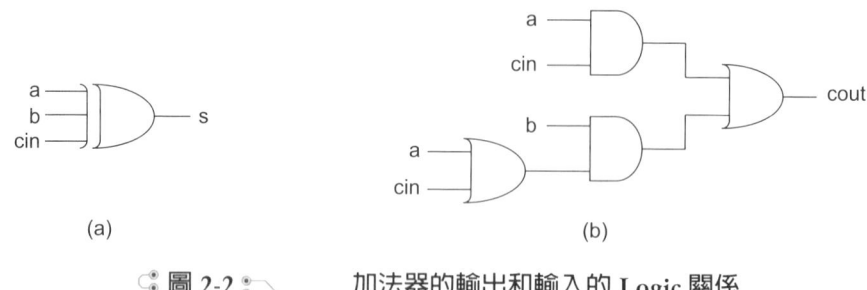

圖 2-2　加法器的輸出和輸入的 Logic 關係

有關 VHDL 電路和 Testbench 的構成，請參考附錄 A 和 B。ModelSim 的 Simulation 步驟，請參考附錄 E。

2-2　VHDL Coding 與 ModelSim Simulation

圖 2-2 簡單加法器電路的 VHDL Coding，如圖 2-3 所示。

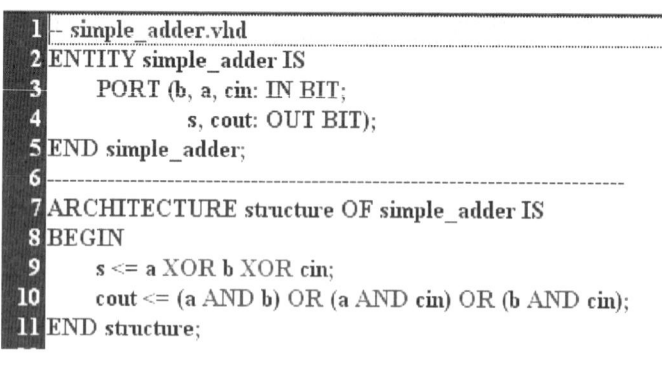

圖 2-3　圖 2-2 加法器電路的 VHDL Coding

模擬測試圖 2-3 加法器電路的 VHDL testbench，如圖 2-4 所示。測試 Truth Table 所需的 Timing Diagram 最簡單方法，是讓訊號的相對時間依次倍增，如 line 19 到 21 所示。line 13 將訊號的開始設定為 '0'，對於測試訊號至關重要。

```vhdl
1  -- testbench for simple_adder
2  ENTITY testbench IS
3  END testbench;
4
5  USE work.ALL;
6
7  ARCHITECTURE stimulus OF testbench IS
8      COMPONENT simple_adder
9          PORT (b, a, cin: IN BIT;
10             s, cout: OUT BIT);
11     END COMPONENT;
12 -- ----------------------------------------------------------
13 SIGNAL b, a, cin: BIT := '0';
14 SIGNAL   s, cout: BIT;
15
16 BEGIN
17     DUT: simple_adder PORT MAP(a, b, cin, s, cout);
18
19     b <= NOT  b  AFTER  50 ns;
20     a <= NOT  a  AFTER 100 ns;
21     cin <= NOT cin AFTER 200 ns;
22
23 END stimulus;
```

圖 2-4　　模擬測試圖 2-3 加法器電路的 VHDL testbench

使用 ModelSim Simulator testbench 測試的結果，如圖 2-5 所示，與圖 2-1 的 Truth Table 相對比，它們完全相同。

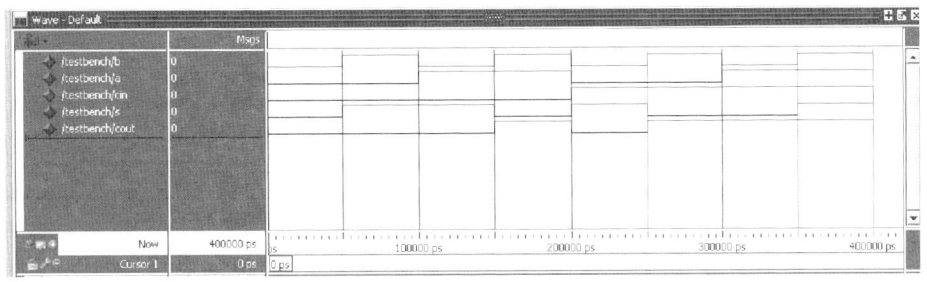

圖 2-5　　模擬測試的結果顯示與 Truth Table 完全相同

2-3　Carry-ripple adder SN74LS283

圖 2-6 為 SN74LS283 4 Bits Carry-ripple adder 的電路圖。

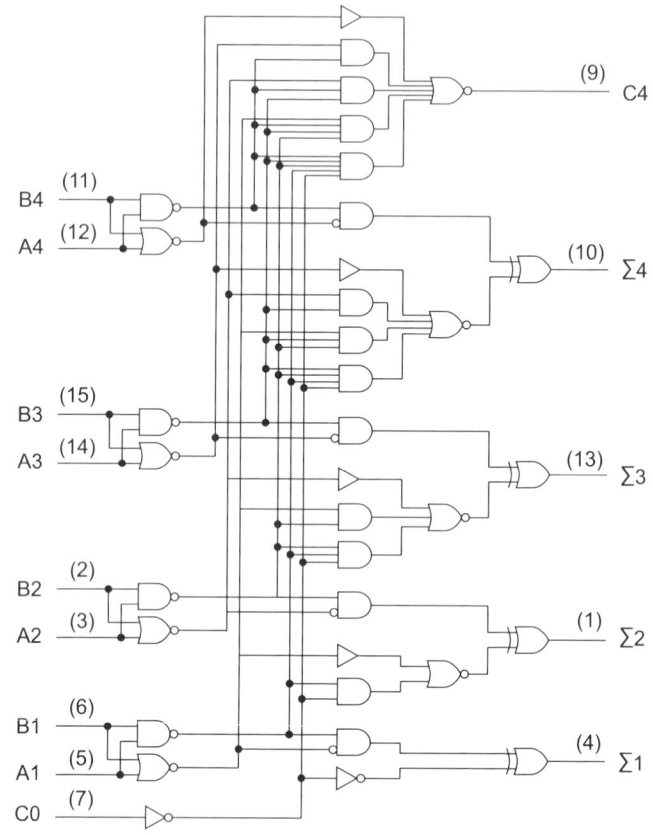

圖 2-6　　SN74LS283 4 Bits Carry-ripple adder 電路

它的輸入為 A1~A4，B1~B4 和 C0。輸出為 ∑1~∑4 和 C4。電路中 Carry Bits C0 ~ C4 呈串連狀態，如圖 2-7 所示，由於 Gate Delay，使得完成加法的速度降低了。以一個 64 bits 加法器為例，假定每個 Gate Delay 為 0.1 ns，則 C4 與 C0 間的 Delay 便是 64 × 0.1 = 6.4 ns！

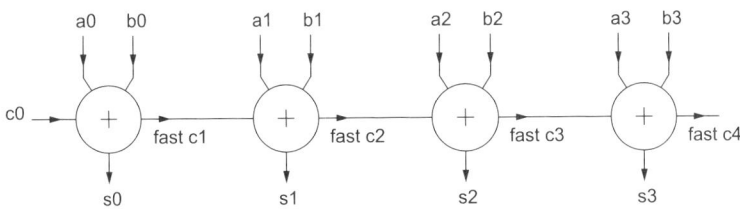

圖 2-7 Carry-ripple adder 電路中 Carry Bits C0~C4 呈串連狀態

2-4　Carry-Lookahead adder

為了免除 Carry 因串連結構而引起的 Delay，Carry-Lookahead adder 電路，對每級的 Carry 採用獨立處理的方式，如圖 2-8 所示。

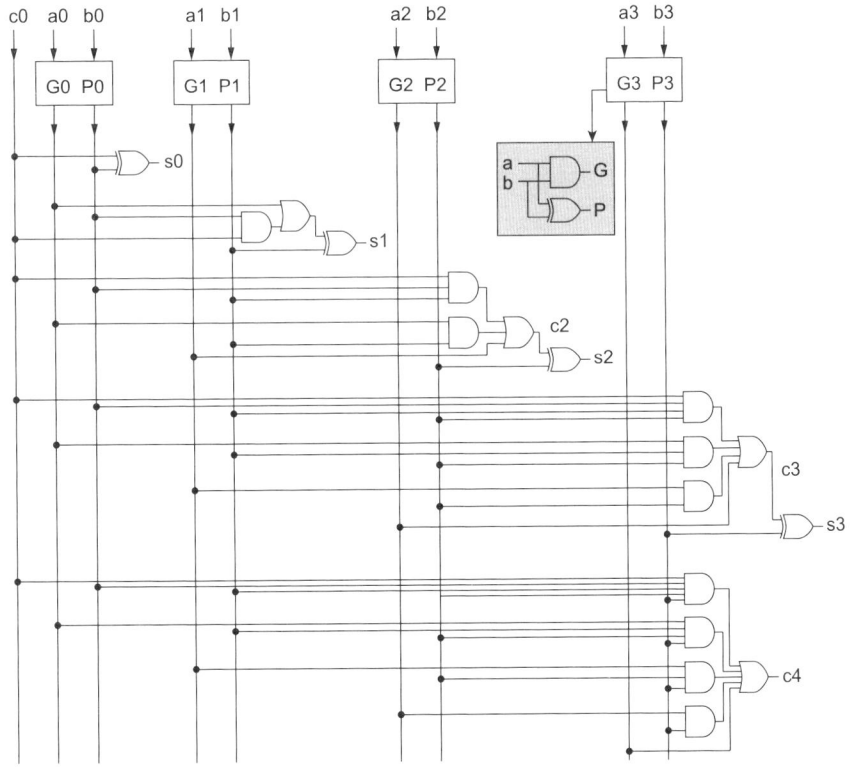

圖 2-8 Carry-Lookahead adder 電路免除了串連結構引起的 Delay

2-5 硬體實作

圖 2-9 為 SN74LS283 4 Bits Carry-ripple adder 的電路圖。它的輸入為 A1~A4，B1~B4 和 C0。輸出為 $\Sigma 1$~$\Sigma 4$ 和 C4。

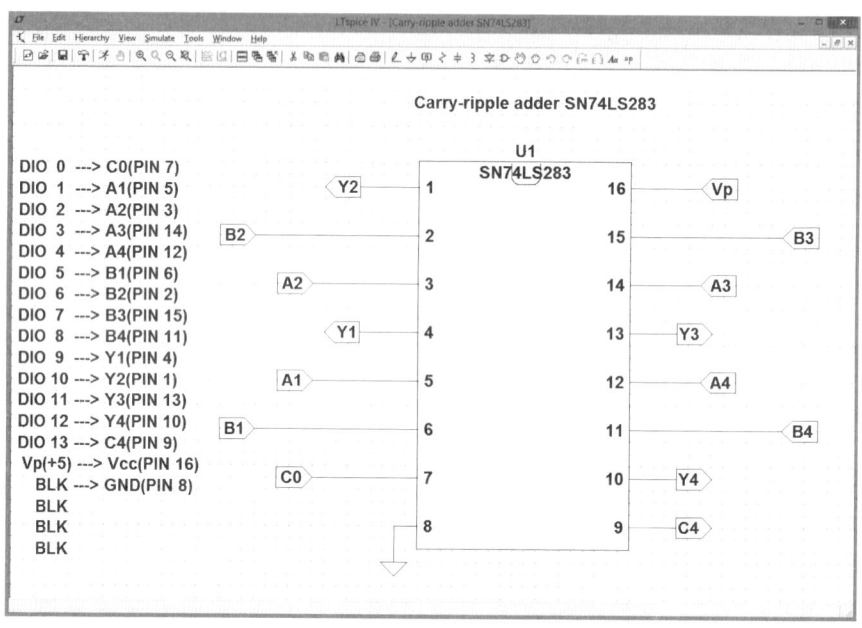

圖 2-9　SN74LS283 電路實體測試圖

簡單的數位電路，如果使用比較低的頻率 (1 KHz)，可以使用不用銲接的**麵包板** (Breadboard) 來完成，如圖 2-10 所示。頻率較高 (10 KHz) 以上或比較複雜的電路，應當把電路裝配在接地良好、使用銲接的電路板上。

第二章　加法器電路的組成和測試　15

◦ 圖 2-10 ◦　　　將電路配置在不用銲接的麵包板

　　將麵包板上之電路，經由接線柱，連接到 Analog Discovery Module，連同裝上了 WaveForms 軟體的 PC，成為多件由軟體操控的硬體儀器。測試數位電路，對於電路的輸入部份可用**數字模式產生器** (Digital Pattern Generator)，輸出部份可用**數位信號分析儀** (Digital Signal Analyzer)，圖 2-11 就是 WaveForms 軟體在 PC 上所顯示的選擇圖形。

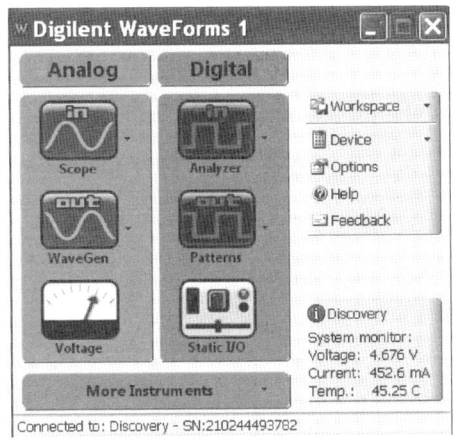

◦ 圖 2-11 ◦　　　WaveForms 軟體在 PC 上所顯示的選擇圖形

點擊 Digital 部份的 out，便可獲得如圖 2-12 的數字模式產生器。有關 Digital Pattern Generator 的設定，請參考附錄 C。

數字模式產生器的設定如下：

圖 2-12 為 SN74LS283 4 Bits Carry-ripple adder 的電路圖，它的輸入為 A1~A4，B1~B4 和 C0。

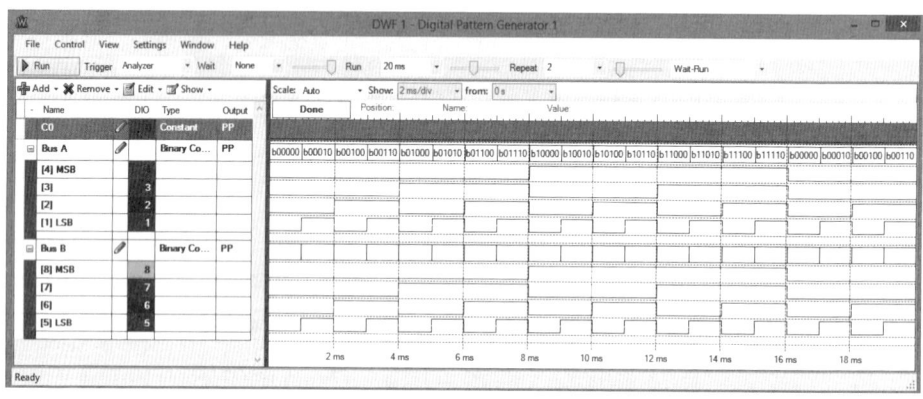

☙ 圖 2-12 ❧　　提供電路輸入訊號的數字模式產生器

1. 將 C0 選用 signal，連接到 Analog Discovery 的 DIO 0 上。
2. 將 A1~A4 選用 Bus A，連接到 Analog Discovery 的 DIO 1~4 上。
3. 令 Bus A 產生 Binary Counter 的 Patterns。
4. 將 B1~B4 選用 Bus B，連接到 Analog Discovery 的 DIO 5~8 上。
5. 令 Bus B 也產生 Binary Counter 的 Patterns。
6. 令 Trigger 為 Analyzer；Wait 為 None；Run 為 20 ms；Repeat 為 1；Wait-Run。

點擊 Digital 部份的 in，便可獲得如圖 2-13 的數位信號分析儀。有關 Logic Analyzer 的設定和使用，請參考附錄 D。數位信號分析儀的設定如下：

圖 2-13 為 SN74LS283 4 Bits Carry-ripple adder 的電路圖，它的輸出為 $\Sigma 1 \sim \Sigma 4$ 和 C4。

1. 將 C4 選用 **signal**，連接到 **Analog Discovery** 的 DIO 13 上。
2. 將 $\Sigma 1 \sim \Sigma 4$ 選用 **Bus Y**，連接到 **Analog Discovery** 的 DIO 9~12 上。

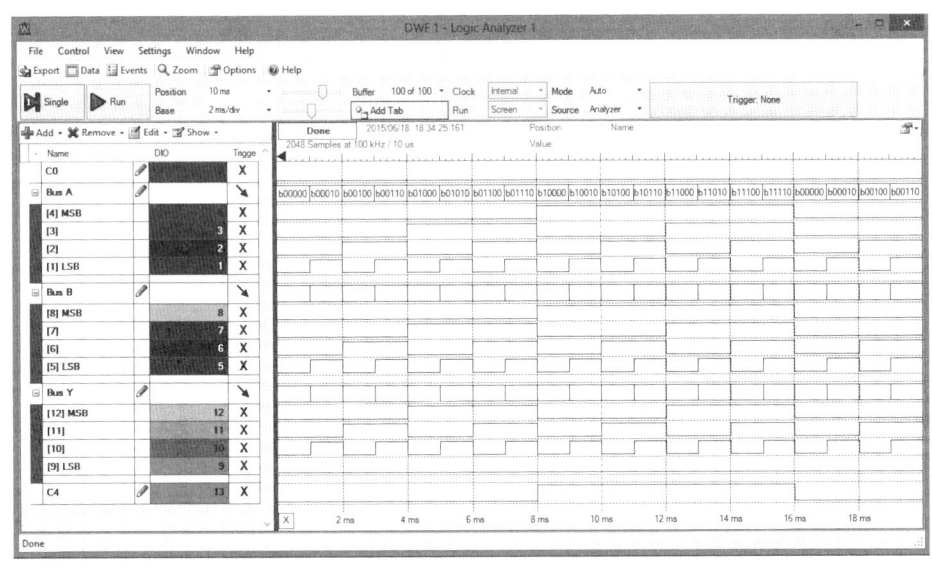

圖 2-13　提供電路記錄輸入訊號的數位信號分析儀

最後，點擊 **Analog** 部份的 **Voltage**，以便 SN74LS283 可獲得如圖 2-14 的 2 個固定 ±5 V 電壓源。TTL IC 使用 V+ 即可。V- OFF。

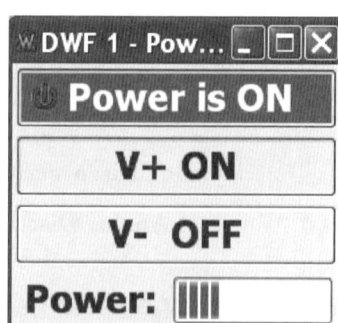

圖 2-14　二個固定 ±5 V 的電壓源

2-6 課外練習

1. 圖 2-6 為 SN74LS283 4 Bits Carry-ripple adder 的電路圖。

   A. 寫出電路的 VHDL 結構編碼。

   B. 用 ModelSim Simulator 來測試證實其結構編碼。

2. 減法器可以將加法器加上 Two's complementer 電路而獲得。試以圖示將 SN74LS283 電路改變成 a ± b 的電路。

3. 圖 2-15 是一個稱為 Incrementer 電路，b 與 a 的關係為：b = a + 1。

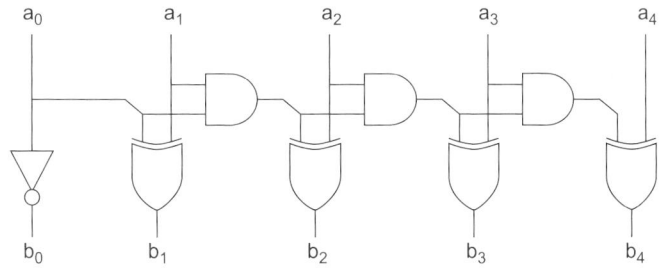

圖 2-15　由 XOR 和 AND gate 組成的 b=a+1 電路

   A. 寫出電路的 VHDL 結構編碼。

   B. 用 ModelSim Simulator 來測試證實其結構編碼。

4. 圖 2-4 的 Line 13: SIGNAL b, a, cin := '0'; 為何對於測試訊號，至關重要？難道不可沒有或改變成其它訊號嗎？

第三章　JKFF 電路的組成和測試

JKFF 電路是**時序邏輯** (Sequential Logic) 的基本元件。組成這個元件的電路種類很多。用途甚廣，諸如 Shift Register、Counter 及各種 Flip Flop 都可以藉 JKFF 來完成。本章為介紹它們的組成、VHDL Coding、Simulation 和硬體測試的方法。

3-1　JKFF 電路的組成

圖 3-1 是 SN74LS112 JKFF 電路的 Gate 組成。

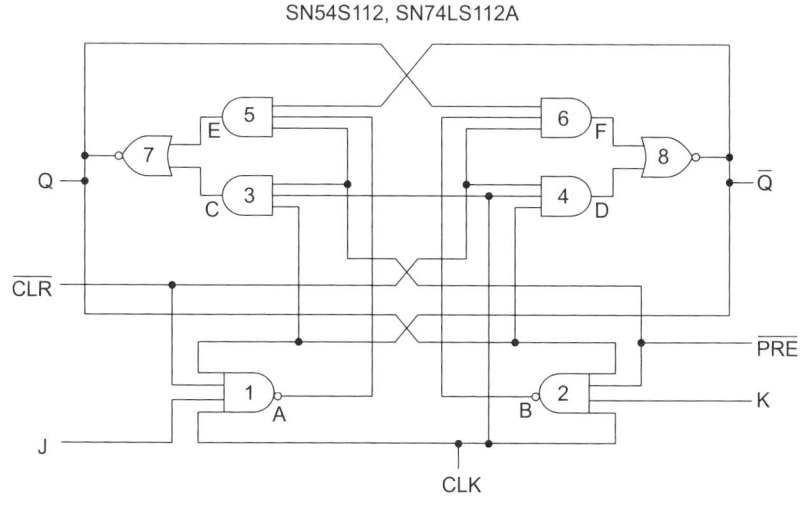

圖 3-1　Gate 組成的 JKFF SN74LS112 電路

它的 Truth Table 如圖 3-2 所示。

INPUTS				OUTPUTS	
$\overline{PRE}$	$\overline{CLR}$	CLK	J	K	Q	$\overline{Q}$
L	H	X	X	X	H	L
H	L	X	X	X	L	H
L	L	X	X	X	H↑	H↑
H	H	↓	L	L	Q0	$\overline{Q0}$
H	H	↓	H	L	H	L
H	H	↓	L	H	L	H
H	H	↓	H	H	TOGGLE	
H	H	H	X	X	Q0	$\overline{Q0}$

圖 3-2　SN74LS112 JKFF 電路的 Truth Table

3-2　VHDL Coding 與 ModelSim Simulation

　　SN74LS112 JKFF 的 VHDL Coding 如圖 3-3 所示。有關 VHDL 電路和 Testbench 的構成，請參考附錄 A 和 B。ModelSim 的 Simulation 步驟請參考附錄 E。

```vhdl
1  LIBRARY IEEE;
2  USE IEEE.STD_LOGIC_1164.ALL;
3  ENTITY SN74LS112 IS
4    PORT( PRE, CLR, J, K, CLK : IN STD_LOGIC;
5          Q, QN : INOUT STD_LOGIC);
6  END ENTITY;
7  ARCHITECTURE structure OF SN74LS112 IS
8  SIGNAL A, B, C, D, E, F: STD_LOGIC;
9  BEGIN
10   U1:  A <= NOT( CLK AND J AND CLR AND Qn);
11   U2:  B <= NOT( CLK AND K AND PRE AND Q);
12   U3:  C <= Qn AND CLK AND PRE;
13   U4:  D <= Q AND CLK AND CLR after 10 ps;
14   U5:  E <= PRE AND A AND Qn after 10 ps;
15   U6:  F <= CLR AND B AND Q;
16   U7:  Q <= C NOR E;
17   U8:  QN <= D NOR F;
18 END structure;
```

圖 3-3　　SN74LS112 JKFF 的 VHDL Coding

　　其中 U4 和 U5 邏輯的尾端加入的 after 10 ps，所造成的 Delay 是產生 positive going edge (↑) 的一種方法。

　　SN74LS112 JKFF 的 testbench 如圖 3-4 所示。其中 PRE 和 CLR 是負性操作訊號，也就是要 '0' 才達到 preset 和 reset 的功能。

```vhdl
-- testbench for SN74LS112
Library IEEE;
USE IEEE.STD_LOGIC_1164.ALL;

ENTITY testbench IS
END testbench;

USE work.all;

ARCHITECTURE stimulus of testbench IS
   component SN74LS112
     PORT( PRE, CLR, J, K, CLK : IN STD_LOGIC;
           Q, QN : INOUT STD_LOGIC);
   end component;
-- Declare testbench's signals
SIGNAL PRE: STD_LOGIC :='1';
SIGNAL CLR: STD_LOGIC :='0';
SIGNAL  J: STD_LOGIC :='0';
SIGNAL  K: STD_LOGIC :='0';
SIGNAL CLK: STD_LOGIC :='0';
SIGNAL  Q: STD_LOGIC;
SIGNAL QN: STD_LOGIC;

BEGIN
   DUT: SN74LS112 PORT MAP(PRE, CLR, J, K, CLK, Q, QN);

-- test signal generation:
   CLR <= '1' AFTER 25 ns;
   CLK <= NOT CLK AFTER  50 ns;
    J <= NOT  J AFTER 100 ns;
    K <= NOT  K AFTER 200 ns;

END stimulus;
```

圖 3-4　SN74LS112 JKFF 的 testbench

使用 ModelSim Simulator testbench 測試的結果如圖 3-5 所示，與圖 3-2 的 Truth Table 相對比，它們完全相同。

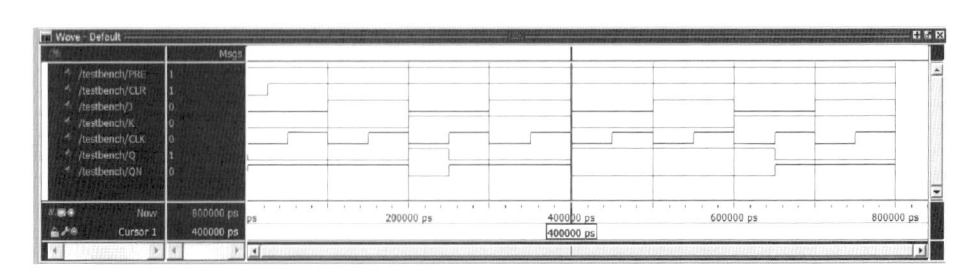

圖 3-5　ModelSim Test Bench 模擬測試的結果

◎ 3-3 硬體實作

圖 3-6 為 SN74LS112 電路實體測試圖。SN74LS112 內有二只 JKFF，本測試僅測試其中的一只，它的輸入標明為 PRE_N、CLR_N、CLK、J 和 K。輸出為 Q 和 Q_N。

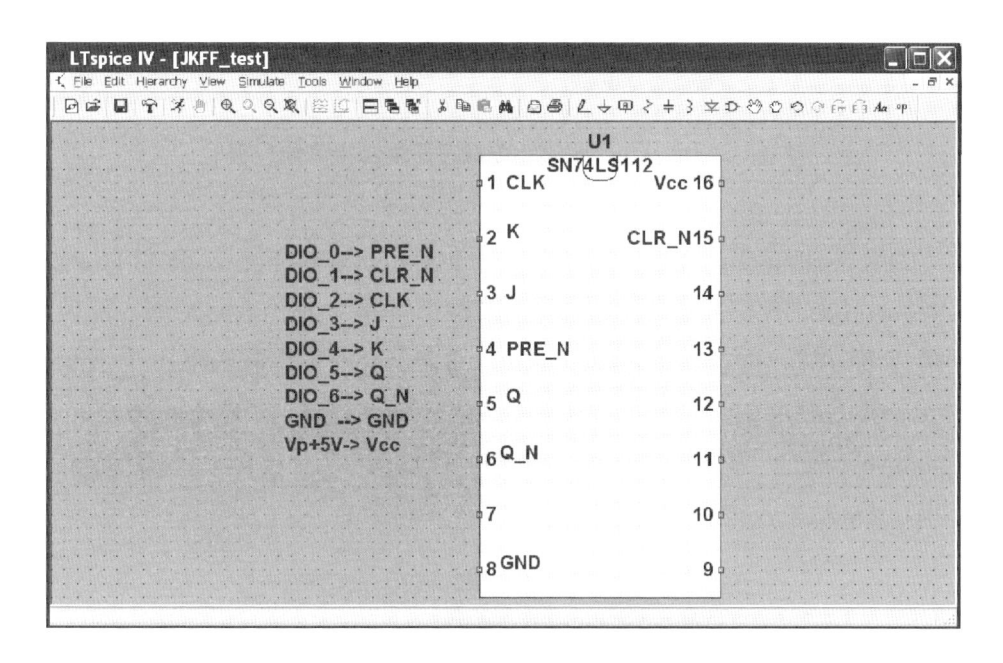

◦ 圖 3-6 ◦　　SN74LS112 電路實體測試圖

簡單的數位電路，如果使用比較低的頻率 (1 KHz)，可以使用不用銲接的**麵包板** (Breadboard) 來完成，如圖 3-7 所示。頻率較高 (10 KHz) 以上或比較複雜的電路，應當把電路裝配在接地良好、使用銲接的電路板上。

◦ 圖 3-7 ◦　　　將電路配置在不用銲接的麵包板上

　　使用 Analog Discovery Module，連同裝上了 WaveForms 軟體的 PC，成為多件由軟體操控的硬體儀器。測試數位電路，對於電路的輸入部份可用**數字模式產生器** (Digital Pattern Generator)，有關 Digital Pattern Generator 的設定，請參考附錄 C。輸出部份可用**數位信號分析儀** (Digital Signal Analyzer)，圖 3-8 就是 WaveForms 軟體在 PC 上所顯示的選擇圖形。有關 Logic Analyzer 的設定和使用，請參考附錄 D。

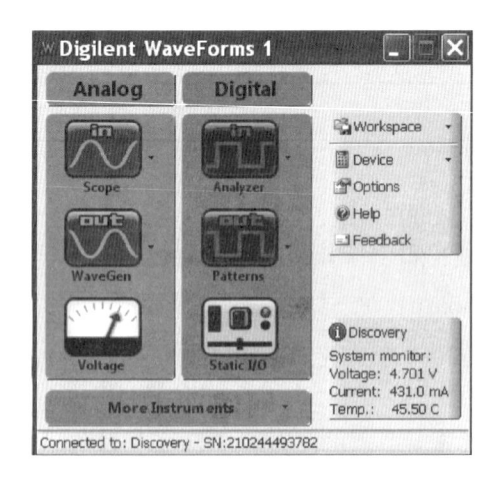

◦ 圖 3-8 ◦　　　WaveForms 軟體在 PC 上所顯示的選擇圖形

點擊 Digital 部份的 out，便可獲得如圖 3-9 的數字模式產生器。

◦ 圖 3-9 ◦　　　提供電路輸入訊號的數字模式產生器

數字模式產生器的設定如下：

圖 3-6 為 SN74LS112 電路實體測試圖。它的輸入為 PRE_N、CLR_N、CLK、J 和 K。由於所有輸入都是獨立的訊號，所以全部以 signal 來處理。其中除了 CLK 需用 CLOCK 來處理外，其它訊號全部用 Custom 來處理[註釋]。

1. PRE_N 訊號連接到 Analog Discovery 的 DIO 0 上。
2. CLR_N 訊號連接到 Analog Discovery 的 DIO 1 上。
3. CLK 連接到 Analog Discovery 的 DIO 2 上。
4. J 訊號連接到 Analog Discovery 的 DIO 3 上。
5. K 訊號連接到 Analog Discovery 的 DIO 4 上。

點擊 Digital 部份的 in，便可獲得如圖 2-13 的數位信號分析儀。

數位信號分析儀的設定如下：

圖 3-6 是 SN74LS112 電路實體測試圖，它的輸出訊號為 Q 和 Q_N。由於觀測電路的輸入和輸出的 Timing 關係，所以把輸入訊號 PRE_N、CLR_N、CLK、J、和 K 全部加入進來，連同：

1. 將 Q 選用 signal，連接到 Analog Discovery 的 DIO 5 上。
2. 將 Q_N 選用 signal，連接到 Analog Discovery 的 DIO 6 上。

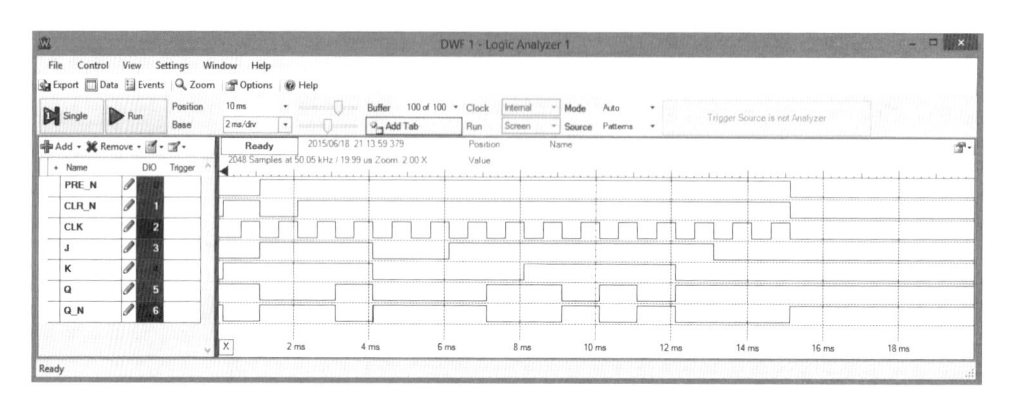

圖 3-10　提供電路輸入訊號的數字模式產生器

最後，點擊 Analog 部份的 Voltage，以便 SN74LS112 可獲得如圖 3-11 的 2 個固定 ±5 V 電壓源。TTL IC 使用 V+ 即可。V- OFF。

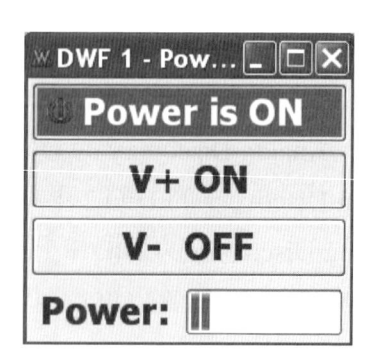

圖 3-11　二個固定 ±5 V 的電壓源

3-4 課外練習

1. D-type Flip Flop 與 D-type Latch 在結構上有何不同？
2. 試將 JKFF 改變成 DFF 和 TFF，請用電路來表明之。
3. 試寫出 JKFF 有關的 Time parameters，並敘述測試的方法。
4. 試述 JKFF 受到 clock skew and slow clock transitions 的影響？
5. 圖 3-3 SN74LS112 JKFF 的 VHDL Coding 是依據 Gates 的結構而成，如果依照 JKFF 的 Truth Table 並且經過 kmap 或邏輯的簡化，可以得到 Behavior 型的 VHDL Coding。請寫出並用 ModelSim Simulator 證實之。

3-5 註釋

訊號 Custom 的設定，以 DIO 0 為例，它是 PRE_N，當其訊號為 '0' 時，Q 被設定為 '1'。如果要測試這個狀態，那麼在開始時設定為 '0'，其它全部為 '1'。設定的步驟為：

1. 點選 **DIO 0**。單擊 **Edit** →**Edit Parameters of "DIO 0"**。如圖 3-12 所示。

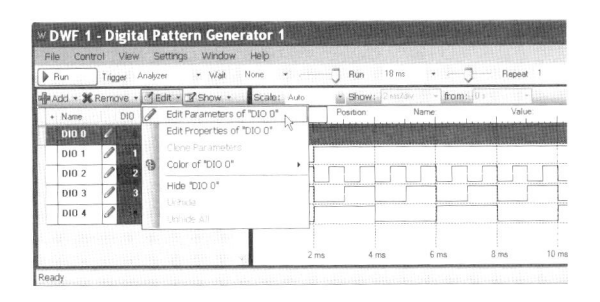

圖 3-12　　Edit Custom 訊號步驟之一

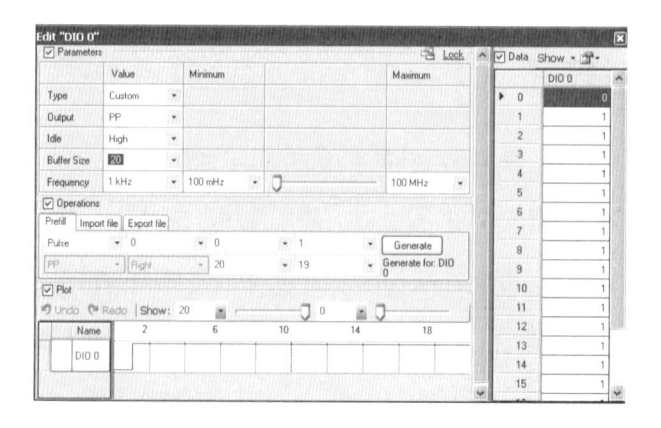

◦ 圖 3-13 ◦　　　Edit Custom 訊號步驟之二

1. 當圖 3-13 出現時，填入各個所需的項目之值。如 Custom, PP, High, 20, 1 KHz 等。
2. 再在這 Edit "DIO 0" 的最下端 2~20 藉 mouse 來選定每個時間內的 '0', 'Z', 或 '1' 之值。同時也顯示在右邊的 Data Buffer 內。
3. 同理可設定 CLR_N，使之在第 1 項為 '0' 外。其它第 0 項和第 2 到 19 項，全部為 '1'。
4. J 和 K 的 Custom 設定，J = K = '1' 的時間應較長，以確定 Q 和 Q_N 能 Toggle。

訊號 Clock 的設定，用 DIO 2 為例，由於 CLK 是整個測試系統的主要訊號。一經選定為 1 KHz，那麼其它訊號也必須選用這個頻率。

第四章　Data Selectors 與 Multiplexers/Demultiplexers

邏輯電路中能歸納成 Data Selectors and Multiplexers/Demultiplexers 一類的，成為代表性的有以下幾種：

SN74LS257　2-line to 1-line Data Selectors/Multiplexers
SN74153　　4-line to 1-line Data Selectors/Multiplexers
SN74LS138　3-line to 8-line Decoder/Demultiplexers
SN74LS148　8-line to 3-line Priority Encoder

本章介紹它們的組成、VHDL Coding、Simulation 和使用 Analog Discovery 來做硬體測試的方法。

4-1　4-line to 1-line Data Selectors/Multiplexers

圖 4-1 是 SN74153 4-line to 1-line Data Selectors/Multiplexers 電路圖，簡稱 MUX，圖的右方是它的簡化方塊圖。

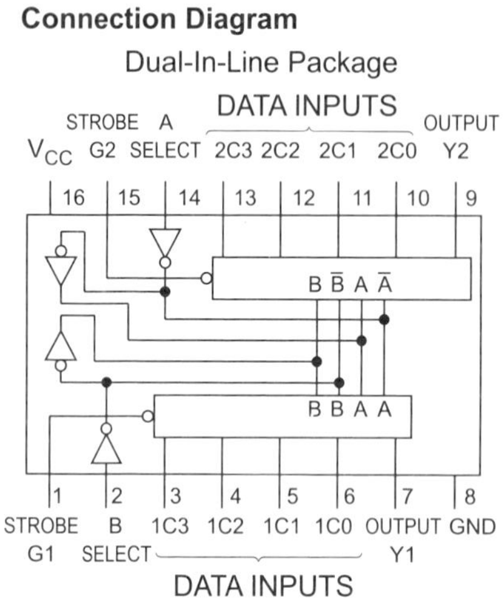

Connection Diagram

Dual-In-Line Package

```
        STROBE  A   DATA INPUTS        OUTPUT
V_CC    G2   SELECT  2C3 2C2 2C1 2C0    Y2
 16      15    14    13  12  11  10      9
```

```
 1      2     3    4   5   6    7      8
STROBE  B   1C3 1C2 1C1 1C0 OUTPUT   GND
  G1      SELECT                Y1
            DATA INPUTS
```

Function Table

Select Inputs		Data Inputs				Strobe	Output
B	A	C0	C1	C2	C3	G	Y
X	X	X	X	X	X	H	L
L	L	L	X	X	X	L	L
L	L	H	X	X	X	L	H
L	H	X	L	X	X	L	L
L	H	X	H	X	X	L	H
H	L	X	X	L	X	L	L
H	L	X	X	H	X	L	H
H	H	X	X	X	L	L	L
H	H	X	X	X	H	L	H

Select inputs A and B are common to both sections.
H = High Level, L = Low Level, S = Don't Care

圖 4-1　　SN74153 電路圖

圖 4-2 為其邏輯型 VHDL code，這種型式的敘述適用於簡單的邏輯電路。有關 VHDL 電路和 Testbench 的構成，請參考附錄 A 和 B。ModelSim 的 Simulation 步驟，請參考附錄 E。

```vhdl
1  LIBRARY ieee;
2  USE ieee.std_logic_1164.all;
3  ----------------------------------------
4  ENTITY mux IS
5      PORT ( a, b, c, d, s0, s1: IN STD_LOGIC;
6             y: OUT STD_LOGIC);
7  END mux;
8  ----------------------------------------
9  ARCHITECTURE pure_logic OF mux IS
10 BEGIN
11     y <= (a AND NOT s1 AND NOT s0) OR
12          (b AND NOT s1 AND s0) OR
13          (c AND s1 AND NOT s0) OR
14          (d AND s1 AND s0);
15 END pure_logic;
```

圖 4-2　　敘述 MUX 的邏輯型 VHDL Code

4-2　3-line to 8-line Decoder/Demultiplexers

圖 4-3 是 SN74LS138：3-line to 8-line Decoder/Demultiplexers 電路圖，簡稱 Encoder，圖的下方是它的簡化方塊圖。

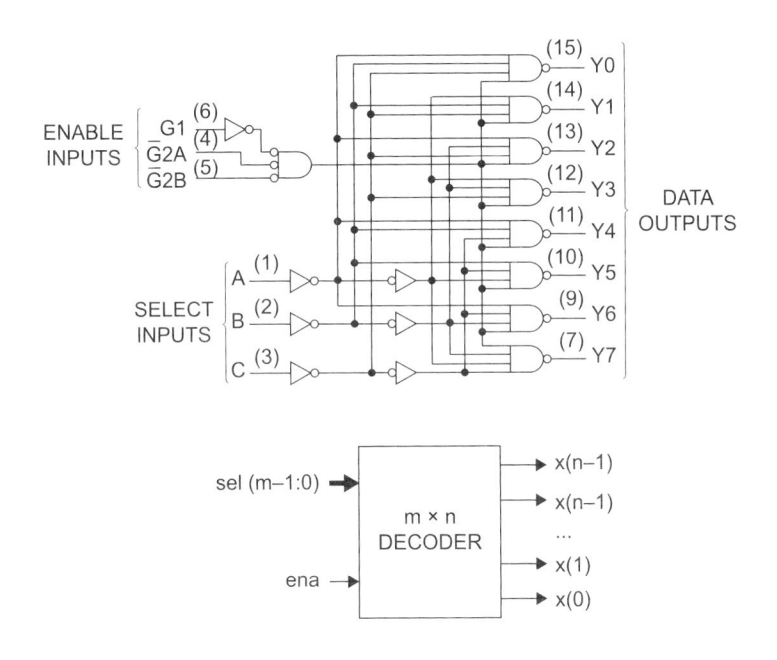

圖 4-3：　ENCODER SN74LS138 電路

圖 4-4 為其**行為型** (Behavior) VHDL code，這種型式的敘述適用於比較複雜的邏輯電路。敘述的過程中看不到任何邏輯的元件。

```vhdl
1  LIBRARY ieee;
2  USE ieee.std_logic_1164.all;
3  ENTITY encoder IS
4      PORT ( x: IN STD_LOGIC_VECTOR (7 DOWNTO 0);
5             y: OUT STD_LOGIC_VECTOR (2 DOWNTO 0));
6  END encoder;
7  --------------------------------------------
8  ARCHITECTURE behavior OF encoder IS
9  BEGIN
10     y <= "000" WHEN x="00000001" ELSE
11         "001" WHEN x="00000010" ELSE
12         "010" WHEN x="00000100" ELSE
13         "011" WHEN x="00001000" ELSE
14         "100" WHEN x="00010000" ELSE
15         "101" WHEN x="00100000" ELSE
16         "110" WHEN x="01000000" ELSE
17         "111" WHEN x="10000000" ELSE
18         "ZZZ";
19 END behavior;
```

圖 4-4　　　SN74LS138 電路的 VHDL 敘述

4-3　MUX 和 DECODER 的測試

　　MUX 和 DECODER 的測試，分別如圖 4-5 和圖 4-7 的 testbench 所示。

第四章 Data Selectors 與 Multiplexers/Demultiplexers

```vhdl
1  -- testbench for MUX
2  LIBRARY ieee;
3  USE ieee.std_logic_1164.all;
4  ------------------------------------------------
5  ENTITY testbench IS
6  END testbench;
7
8  USE work.ALL;
9  ------------------------------------------------
10 ARCHITECTURE stimulus OF testbench IS
11     COMPONENT mux
12     PORT ( a, b, c, d, s0, s1: IN STD_LOGIC;
13            y: OUT STD_LOGIC);
14     END COMPONENT;
15 -- ---------------------------------------------
16 SIGNAL a, b, c, d, s0, s1: STD_LOGIC := '0';
17 SIGNAL y: STD_LOGIC;
18
19 BEGIN
20     DUT: MUX PORT MAP(a, b, c, d, s0, s1, y);
21     a <= '1'; b <= '0';
22     c <= NOT c after 50 US;
23     d <= NOT d after 100 US;
24     s0 <= NOT s0 after 0.5 mS;
25     s1 <= NOT s1 after 1 mS;
26 END stimulus;
```

圖 4-5　測試圖 4-2 MUX 的 testbench code

輸入訊號設定：a 為 '1'，b 為 '0'，c 為 100 μS 週期的方波，d 為 200 μS 週期的方波。選擇訊號：s0, s1 的方波週期為 100 mS 和 200 μS。

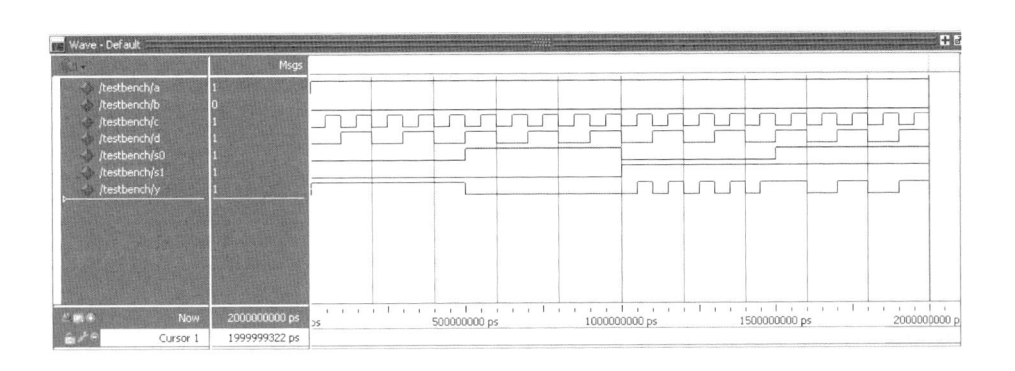

圖 4-6　Testbench 測試 MUX 的結果

```
1  -- testbench for encoder
2  LIBRARY ieee;
3  USE ieee.std_logic_1164.all;
4  ----------------------------------------------------------
5  ENTITY testbench IS
6  END testbench;
7
8  USE work.ALL;
9  ----------------------------------------------------------
10 ARCHITECTURE stimulus OF testbench IS
11     COMPONENT encoder
12         PORT ( x: IN STD_LOGIC_VECTOR (7 DOWNTO 0);
13                y: OUT STD_LOGIC_VECTOR (2 DOWNTO 0));
14     END COMPONENT;
15 -- ---------------------------------------------------------
16 SIGNAL x: STD_LOGIC_VECTOR (7 DOWNTO 0) := "00000000";
17 SIGNAL y: STD_LOGIC_VECTOR (2 DOWNTO 0);
18
19 BEGIN
20     DUT: encoder PORT MAP(x, y);
21
22     x <= "00000001" after 100 ms, "00000010" after 200 ms,
23          "00000100" after 300 ms, "00001000" after 400 ms,
24          "00010000" after 500 ms, "00100000" after 600 ms,
25          "01000000" after 700 ms, "10000000" after 800 ms;
26 END stimulus;
```

圖 4-7　測試圖 4-3 ENCODER 的 Testbench code

輸入訊號 x 的設定用 STD_LOGIC_VECTOR (7 downto 0) 來顯示。

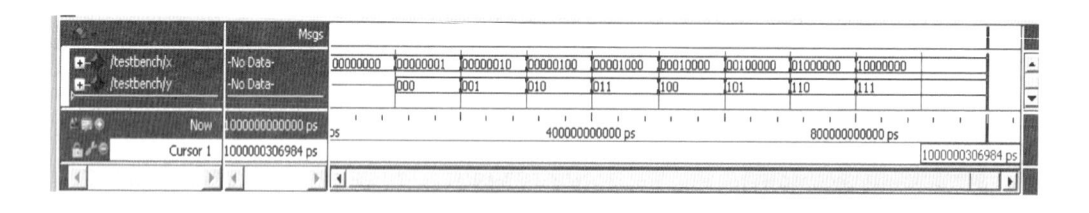

圖 4-8　Testbench 測試 ENCODER 的結果

4-4　MUX 電路在 FPGA 中的用途

第一章裡有使用 NAND gate 來連接成其它 Gates 的敘述。FPGA 電路裡更有使用 4 to 1 MUX 為基礎來連接成其它 Gates 的，而且似乎更為簡單。如圖 4-9：

1. 將 a <= '0', b <='0', c <= '0', d <= '1' 則 s0, s1 和 y 便成 AND gate。
2. 將 a <= '1', b <='1', c <= '1', d <= '0' 則 s0, s1 和 y 便成 NAND gate。
3. 將 a <= '0', b <='1', c <= '1', d <= '1' 則 s0, s1 和 y 便成 OR gate。
4. 將 a <= '1', b <='0', c <= '0', d <= '0' 則 s0, s1 和 y 便成 NOR gate。
5. 將 a <= '0', b <='1', c <= '1', d <= '0' 則 s0, s1 和 y 便成 XOR gate。
6. 將 a <= '1', b <='0', c <= '0', d <= '1' 則 s0, s1 和 y 便成 XNOR gate。

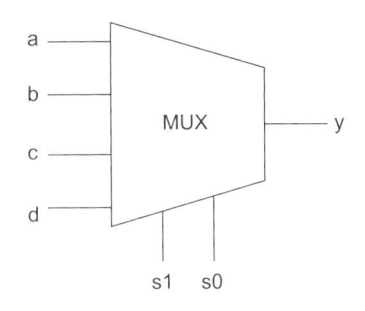

圖 4-9　4 to 1 MUX 示意圖

使用 4 to 1 MUX 來設定成其它 Gate 除了簡單的優點外，還有所有 Gates 的 Delay 完全相同。

4-5 硬體實作

圖 4-10 為 SN74153 Dual 4-Line to 1-Line Data Selectors 的電路圖。它的輸入為 C0~C3，s0~s1 和 EN_N。輸出為 Y。這一個實作是做 4.4 節將 MUX 轉變為 2-input Gates 的測試。

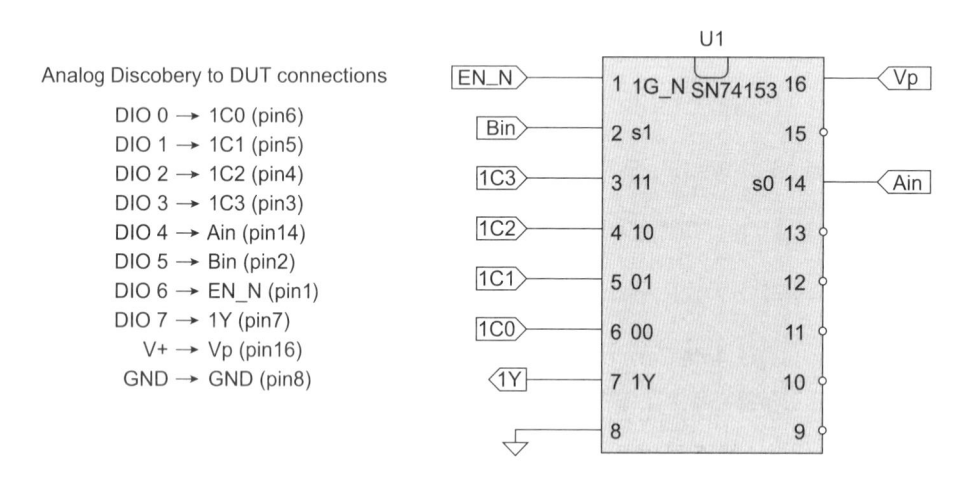

```
Analog Discobery to DUT connections
    DIO 0 → 1C0 (pin6)
    DIO 1 → 1C1 (pin5)
    DIO 2 → 1C2 (pin4)
    DIO 3 → 1C3 (pin3)
    DIO 4 → Ain (pin14)
    DIO 5 → Bin (pin2)
    DIO 6 → EN_N (pin1)
    DIO 7 → 1Y (pin7)
       V+ → Vp (pin16)
      GND → GND (pin8)
```

圖 4-10　SN74153 MUX 轉變為 2-input Gates 的測試

簡單的數位電路，如果使用比較低的頻率 (1 KHz)，可以使用不用銲接的**麵包板** (Breadboard) 來完成，如圖 4-11 所示。頻率較高 (100 KHz) 以上或比較複雜的電路，應當把電路裝配在接地良好，使用銲接的電路板上。

第四章　Data Selectors 與 Multiplexers/Demultiplexers　39

圖 4-11　將電路配置在不用銲接的麵包板上

使用 Analog Discovery Module，連同裝上了 WaveForms 軟體的 PC，成為多件由軟體操控的硬體儀器。測試數位電路，對於電路的輸入部份可用**數字模式產生器** (Digital Pattern Generator)，輸出部份可用**數位信號分析儀** (Digital Signal Analyzer)，圖 4-12 就是 WaveForms 軟體在 PC 上所顯示的選擇圖形。

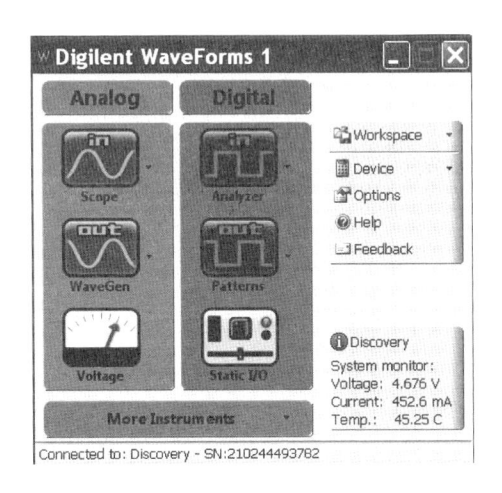

圖 4-12　WaveForms 軟體在 PC 上所顯示的選擇圖形

點擊 Digital 部份的 out，便可獲得如圖 4-14 的數字模式產生器。

數字模式產生器的設定如下：

圖 4-10 為 SN74153 Dual 4-Line to 1-Line Data Selectors 的電路圖。它的輸入為 C0~C3，s0~s1 和 EN_N。輸出為 Y。

1. 將 C0~C3 選用 Bus A，連接到 Analog Discovery 的 DIO 0~3 上。令 Bus A 產生 Custom 的 Patterns 設定，如圖 4-13 所示。
2. 將 s0~s1 選用 signal/clock，s0 的頻率為 2 KHz。s1 的頻率為 1 KHz。
3. EN_N 為 Chip Enable，'0' 執行，'1' 為停止執行。

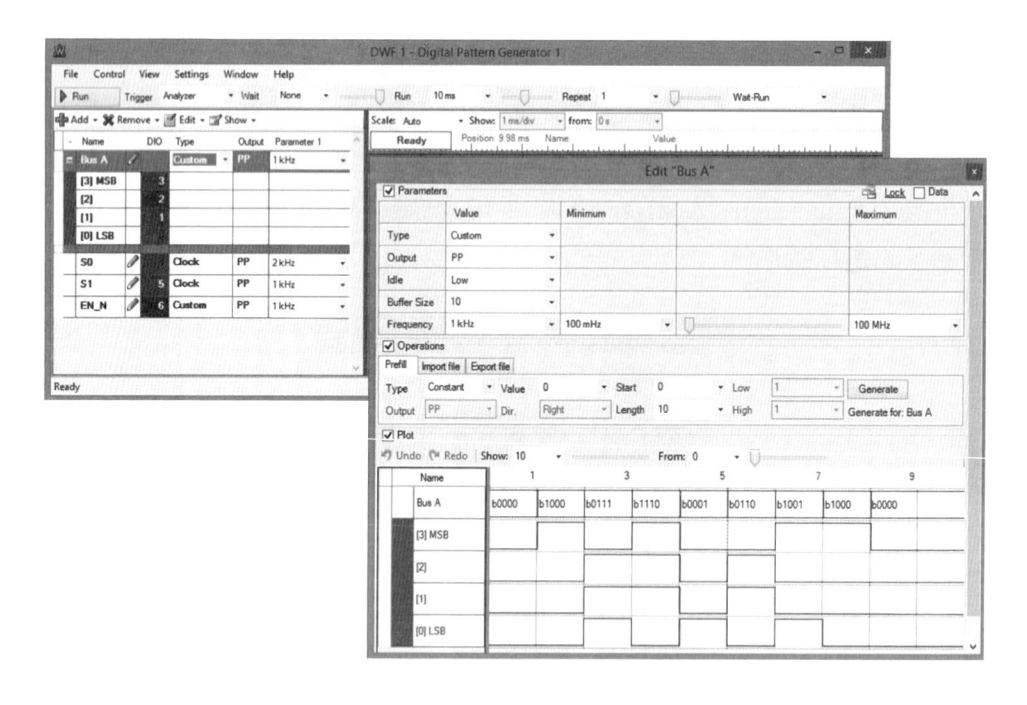

◎ 圖 4-13 ◎　　　C0～C3 BUS A 的設定

結果 Digital Pattern Generator 1 的波形當如圖 4-14 所示。

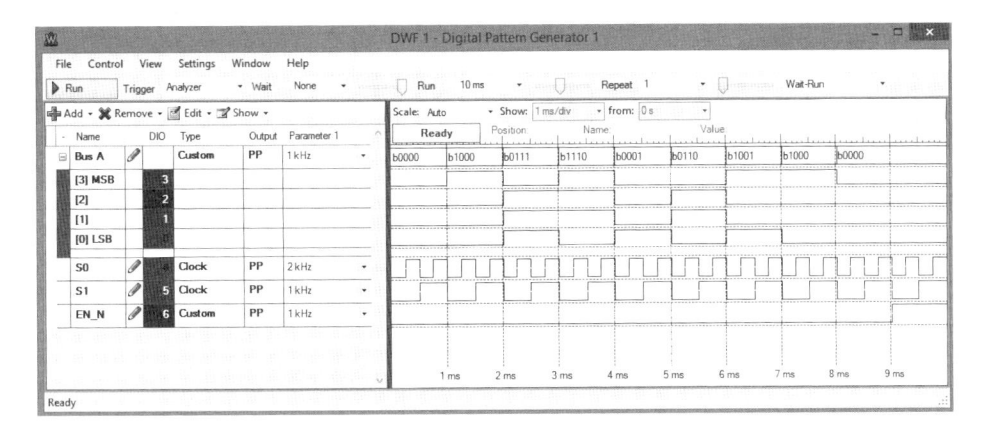

第四章　Data Selectors 與 Multiplexers/Demultiplexers

○ 圖 4-14 ○　　Digital Pattern Generator 1 的波形

點擊 **Digital** 部份的 **in**，便可獲得如圖 4-15 的數位信號分析儀。本實作的輸出訊號只有一個 DIO 7 = Y。

數位信號分析儀的設定如下：

圖 4-15 數位信號分析儀內加入了 Digital Pattern Generator 1 的波形，為的是容易鑑別 I/O 訊號間的 Timing 關係。

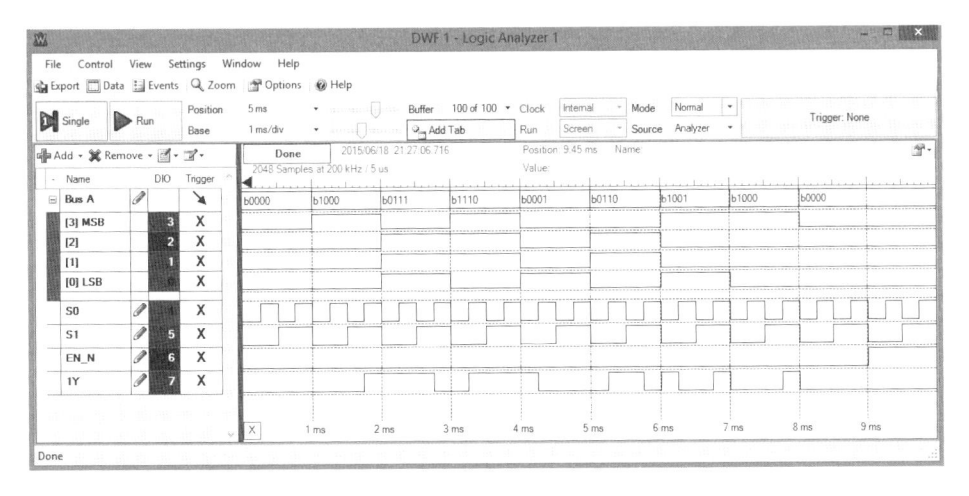

○ 圖 4-15 ○　　數位信號分析儀顯示測試的結果

執行 Logic Analyzer 的測試工作，除了要準備以上的安排及設定之外，還有 SN74153 TTL 所需的 V+ 5 V 電源。點擊圖 4-12 WaveForms 的 **Voltage** 當可獲得電源如下圖所示。選用 V+ ON 及 Power is ON。

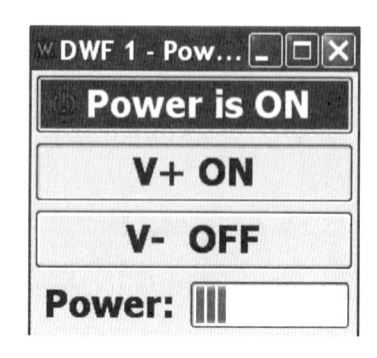

圖 4-16　　TTL SN74153 只需用 V+ 5 V 電源

4-6 課外練習

1. 試寫出 74LS138: 3-line to 8-line Decoder/Demultiplexers 的 VHDL Behavior 模式，並且使用 ModelSim Simulator 的 Testbench 來測試之。

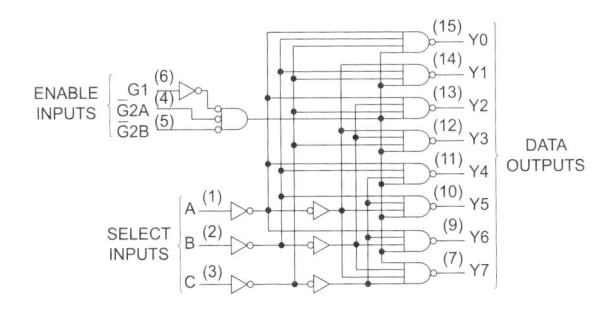

INPUTS					OUTPUTS							
ENABLE		SELECT											
G1	$\overline{G2}$ *	C	B	A	Y0	Y1	Y2	Y3	Y4	Y5	Y6	Y7
X	H	X	X	X	H	H	H	H	H	H	H	H
L	X	X	X	X	H	H	H	H	H	H	H	H
H	L	L	L	L	L	H	H	H	H	H	H	H
H	L	L	L	H	H	L	H	H	H	H	H	H
H	L	L	H	L	H	H	L	H	H	H	H	H
H	L	L	H	H	H	H	H	L	H	H	H	H
H	L	H	L	L	H	H	H	H	L	H	H	H
H	L	H	L	H	H	H	H	H	H	L	H	H
H	L	H	H	L	H	H	H	H	H	H	L	H
H	L	H	H	H	H	H	H	H	H	H	H	L

2. 試用 Analog Discovery 來測試 SN74LS148: 8-line to 3-line Priority Encoder。注意：輸入 8-line 和輸出 3-line 請用 Bus 來構成。

INPUTS									OUTPUTS			
EI	0	1	2	3	4	5	6	7	A2	A1	A0	GS	EO
H	X	X	X	X	X	X	X	X	H	H	H	H	H
L	H	H	H	H	H	H	H	H	H	H	H	H	L
L	X	X	X	X	X	X	X	L	L	L	L	L	H
L	X	X	X	X	X	X	L	H	L	L	H	L	H
L	X	X	X	X	X	L	H	H	L	H	L	L	H
L	X	X	X	X	L	H	H	H	L	H	H	L	H
L	X	X	X	L	H	H	H	H	H	L	L	L	H
L	X	X	L	H	H	H	H	H	H	L	H	L	H
L	X	L	H	H	H	H	H	H	H	H	L	L	H
L	L	H	H	H	H	H	H	H	H	H	H	L	H

3. 試改變 74LS257 Quad 2-line to 1-line data selectors/multiplexers 電路內部的連接成為 Dual 4-line to 2-line data。(允許加減 Gates)

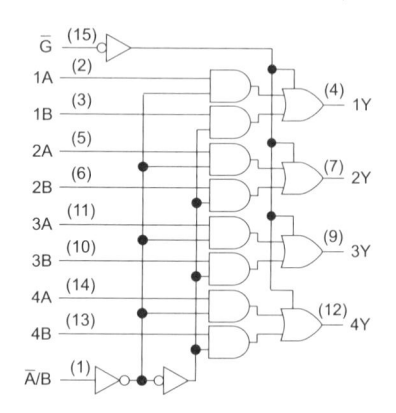

4. 試改變 74LS148 電路外部的連接，使成為 4-line to 2-line data selectors / multiplexers。

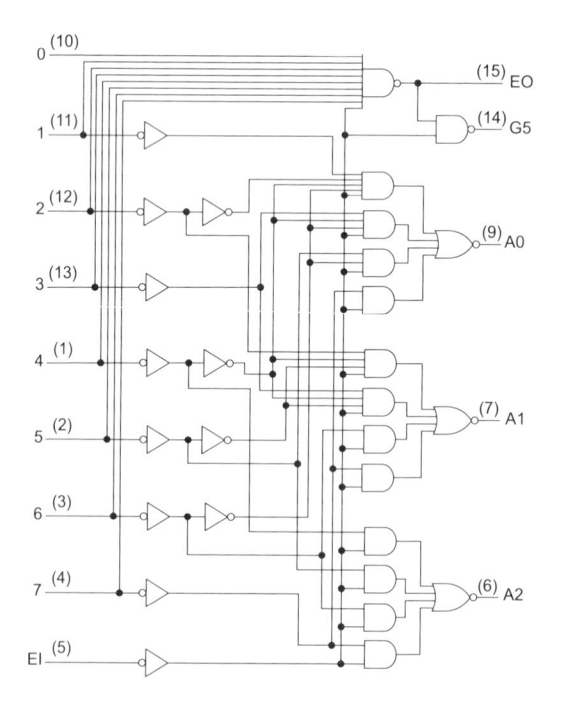

第五章 移位寄存器

Shift Register 是數位系統中主要的子系統之一，由數位元件 Flip Flop 所組成。系統中訊號的並聯和串連的相互變換，就是靠**移位寄存器** (Shift Register) 來完成，是通訊和網路系統中不可缺少的組件。

5-1 DFF 組成的移位寄存器

標準型的 Shift Register 如圖 5-1 所示，它是由 4 只 DFF 組成的 4 bit Shift Register。這個 Shift Register 的 A1~A4 並聯輸入 Data 可以用一個 Positive going 脈波直接來到 Q1~Q4 並聯輸出。Q1~Q4 又可用 4 個從 Clockin 的 Clock 由左到右由 SerialOut 串連輸出。串連的訊號也可

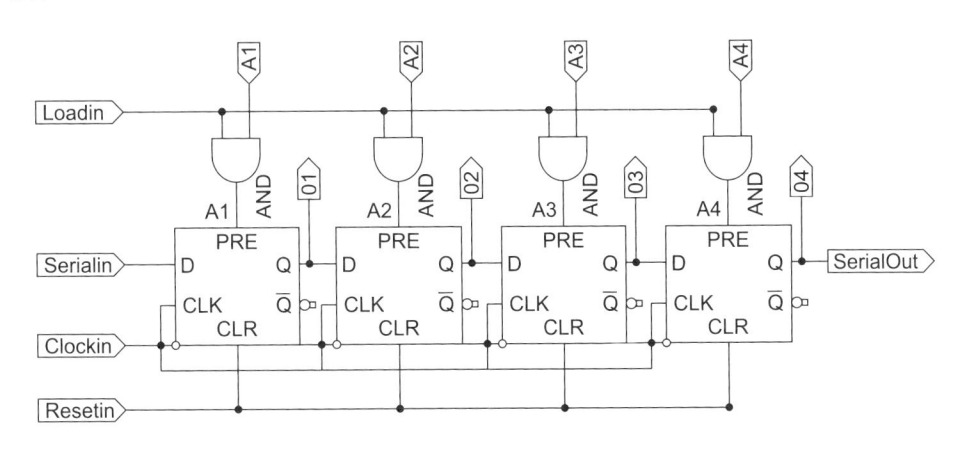

圖 5-1　DFF 組成的 4 bit Shift Register

以從 Serialin 輸入，由 Clock 控制，逐漸地由 SerialOut 由左到右，串連輸出。Positive going 脈波輸入到 Resetin 後，將使 Q1~Q4 重新設定成 "0000"。

5-2 使用 VHDL 來描述電路的結構和電路的模擬測試

對於較複雜的電路，除了使用硬體結構來描述電路，還可以用電路的 I/O **行為** (Behavior) 模式，來對圖 5-1 做 VHDL 的電路的行為描述，如圖 5-2 所示。有關 VHDL 電路和 Testbench 的構成，請參考附錄 A 和 B。ModelSim 的 Simulation 步驟，請參考附錄 E。

```vhdl
1  ---- shift_register
2  LIBRARY ieee;
3  USE ieee.std_logic_1164.all;
4  --------------------------------------------
5  ENTITY shift_register IS
6      PORT ( Serialin, Clockin, Resetin, Loadin: IN STD_LOGIC;
7             A: IN STD_LOGIC_VECTOR(3 DOWNTO 0);
8             Serialout: OUT STD_LOGIC);
9  END shift_register;
10 --------------------------------------------
11 ARCHITECTURE behavior OF shift_register IS
12     SIGNAL internal: STD_LOGIC_VECTOR (3 DOWNTO 0) := "0000";
13 BEGIN
14     PROCESS (Clockin, Resetin, Loadin)
15     BEGIN
16         IF (Resetin='1') THEN
17             internal <= (OTHERS => '0');
18         ELSIF (Clockin'EVENT AND Clockin='1') THEN
19             internal <= Serialin & internal(3 DOWNTO 1);
20         ELSIF (Loadin='1') THEN
21             internal <= A;
22         END IF;
23     END PROCESS;
24 Serialout <= internal(0);
25 END behavior;
26 --------------------------------------------
```

圖 5-2　用 VHDL 對圖 5-1 電路的行為描述

```vhdl
1  ---- Testbench for shif_tregister
2  LIBRARY ieee;
3  USE ieee.std_logic_1164.all;
4  ------------------------------------------
5  ENTITY testbench IS
6  END testbench;
7  ------------------------------------------
8  USE work.ALL;
9  ------------------------------------------
10 ARCHITECTURE stimulus OF testbench IS
11   COMPONENT shift_register
12   PORT ( Serialin, Clockin, Resetin, Loadin: IN STD_LOGIC;
13            A: IN STD_LOGIC_VECTOR(3 DOWNTO 0);
14            Serialout: OUT STD_LOGIC);
15   END COMPONENT;
16 ------------------------------------------------------------
17 SIGNAL  Serialin: STD_LOGIC := '0';
18 SIGNAL   Clockin: STD_LOGIC := '0';
19 SIGNAL   Resetin: STD_LOGIC := '0';
20 SIGNAL   Loadin: STD_LOGIC := '0';
21 SIGNAL        A: STD_LOGIC_VECTOR (3 DOWNTO 0) := "0000" ;
22 SIGNAL Serialout: STD_LOGIC := '0';
23
24 BEGIN
25    DUT: shift_register PORT MAP (Serialin, Clockin, Resetin, Loadin, A, Serialout);
26
27    Clockin <= NOT Clockin AFTER 50 ns;
28    Resetin <= '0';
29    Serialin <= '1' AFTER 100 ns, '0' AFTER 200 ns;
30    A <= "1111" AFTER 200 ns ;
31    Loadin <= '1' AFTER 600 ns, '0'  AFTER 700 ns;
32 END stimulus;
```

圖 5-3　用來測試圖 5-2 的 VHDL TestBench

　　圖 5-3 的 VHDL testbench 是用來測試圖 5-2 shift_register.vhd 的檔案。測試的結果，顯示在圖 5-4 的 Timing Diagram 上。

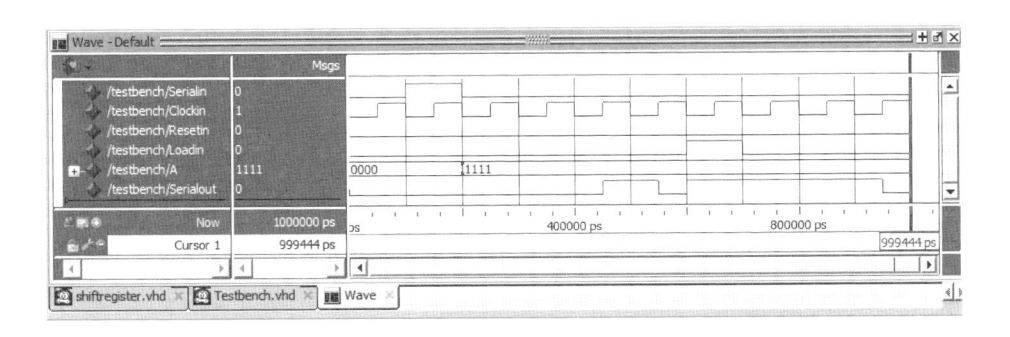

圖 5-4　VHDL TestBench 測試的結果

5-3 雙向 Shift Register SN74LS194 的介紹

雙向 Shift Register SN74LS194 是由 4 只 RSFF 連接成 DFF 組合而成。雙向指的是 Serial data 可以由左到右，也就是 Q1~Q4 串連輸出。也可以由右到左，那便是 Q4~Q1 串連輸出。為了達到雙向的功能，必須加入 Mode Control 控制輸入和輸出電路的連接，圖 5-5 中除了 FF 之外的 Gates 所組成的 concurrent 電路。完成的多項功能，如圖 5-6 表所示。

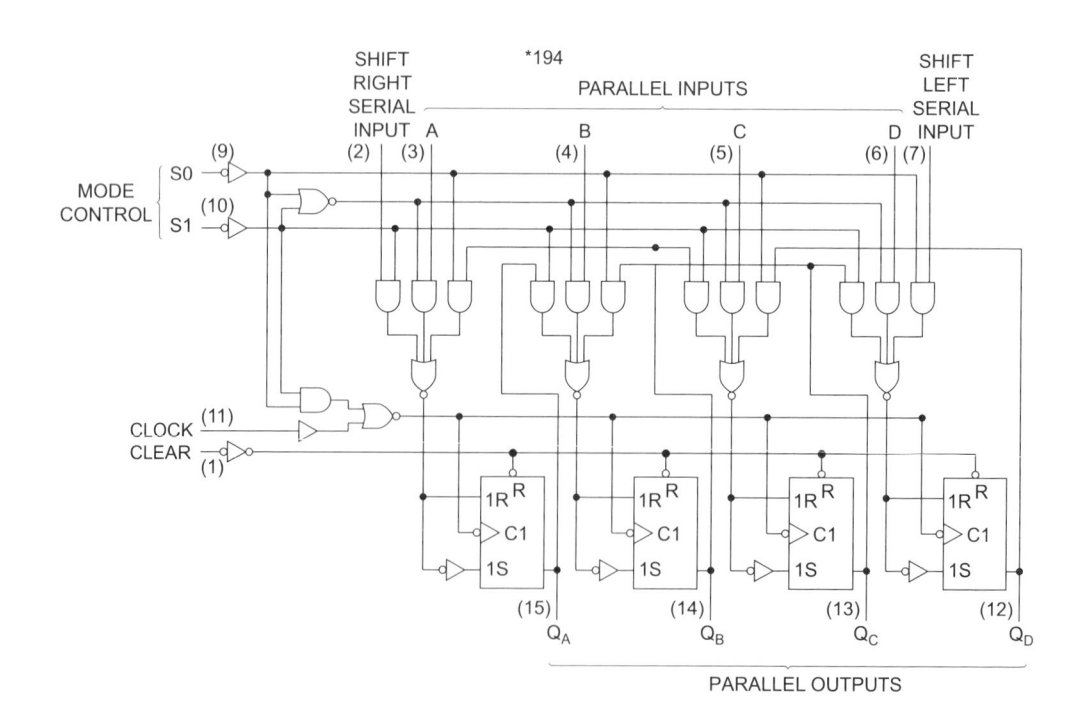

圖 5-5　雙向 Shift Register SN74LS194 電路圖

CLEAR	MODE		CLOCK	SERIAL		PARALLEL				OUTPUTS			
	S1	S0		LEFT	RIGHT	A	B	C	D	QA	QB	QC	QD
L	X	X	X	X	X	X	X	X	X	L	L	L	L
H	X	X	L	X	X	X	X	X	X	QA0	QB0	QC0	QD0
H	H	H	↑	X	X	a	b	c	d	a	b	c	d
H	L	H	↑	X	H	X	X	X	X	H	QAn	QBn	QCn
H	L	H	↑	X	L	X	X	X	X	L	QAn	QBn	QCn
H	H	L	↑	H	X	X	X	X	X	QBn	QCn	QDn	H
H	H	L	↑	L	X	X	X	X	X	QBn	QCn	QDn	L
H	L	L	X	X	X	X	X	X	X	QA0	QB0	QC0	QD0

圖 5-6　雙向 Shift Register SN74LS194 的功能表

圖 5-6 的雙向 Shift Register SN74LS194 功能表，其中的 Mode 控制：

1. 同步並聯**裝載** (Load) 是將 4 個 Bits 的數據 A, B, C, D 當 S0 和 S1 都是 '1' 時完成。同時在 CLK 訊號輸入的上升沿出現時，在 Q0～Q3 輸出。裝載期間，串連數據的輸出停止。

2. 右移是在 s0 = '1'，s1 = '0'，clk 訊號脈衝上升沿時開始，串連數據進入右移模式的數據輸入端。而當 s0 = '0'，s1 = '1'，CLK 訊號脈衝上升沿時，串連數據進入左移模式的數據輸入端。而當 s0 = s1 = '0'，移位儲存器進入靜止狀態。

圖 5-6 的功能表，也就是 74LS194 的 Truth Table，它是電路設計的目標和依據。使用該表不但能寫出 VHDL Behavior 的模式，而且能寫出測試用的 Testbench。圖 5-7 廠商所提供的 Timing Diagram，是很好的參考資料。在最後一節練習裡，將要求讀者依據該表來寫出雙向 Shift Register 的 VHDL Device file，和測試該檔案的 VHDL Testbench。

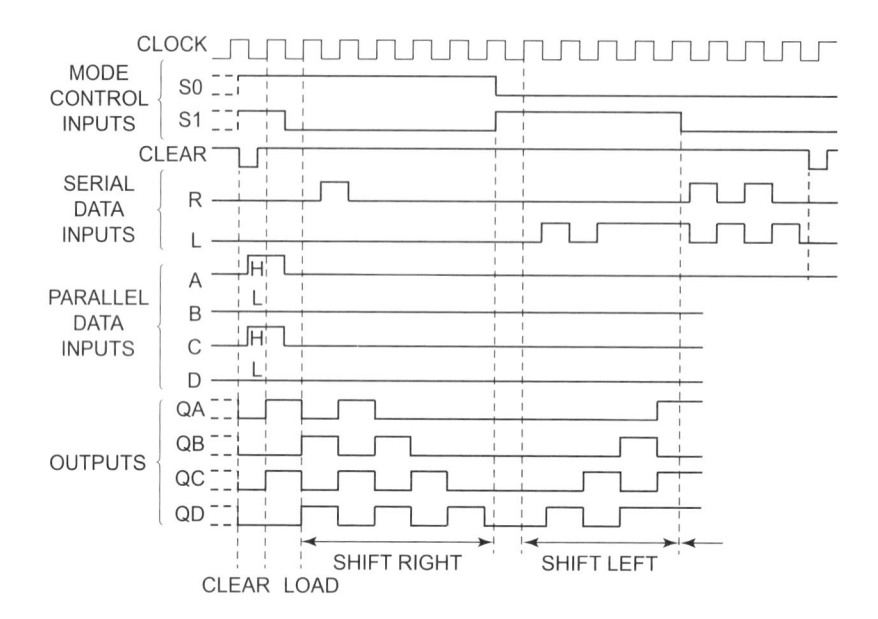

圖 5-7　廠商所提供的 74LS194 Timing Diagram

○ 5-4　硬體實作

圖 5-8 為 Analog Discovery 與 74LS194 的連線圖。

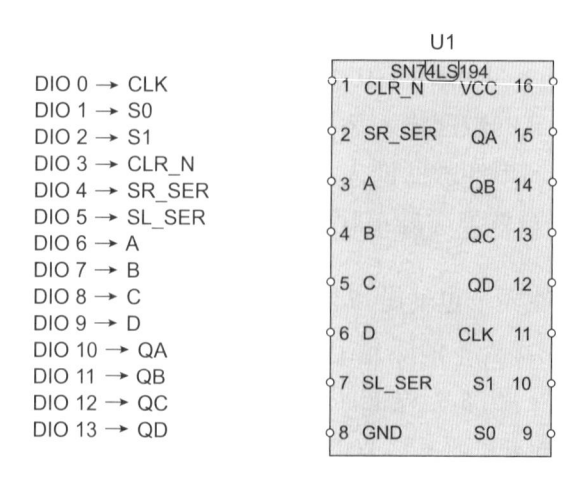

DIO 0 → CLK
DIO 1 → S0
DIO 2 → S1
DIO 3 → CLR_N
DIO 4 → SR_SER
DIO 5 → SL_SER
DIO 6 → A
DIO 7 → B
DIO 8 → C
DIO 9 → D
DIO 10 → QA
DIO 11 → QB
DIO 12 → QC
DIO 13 → QD

U1 SN74LS194
1　CLR_N　　VCC　16
2　SR_SER　　QA　15
3　A　　　　　QB　14
4　B　　　　　QC　13
5　C　　　　　QD　12
6　D　　　　　CLK　11
7　SL_SER　　S1　10
8　GND　　　 S0　 9

圖 5-8　Analog Discovery 與 74LS194 連線圖

簡單的數位電路，如果使用比較低的頻率 (1 KHz)，可以使用不用銲接的**麵包板** (Breadboard) 來完成，如圖 5-9 所示。頻率較高 (10 KHz) 以上或比較複雜的電路，應當把電路裝配在接地良好、使用銲接的電路板上。

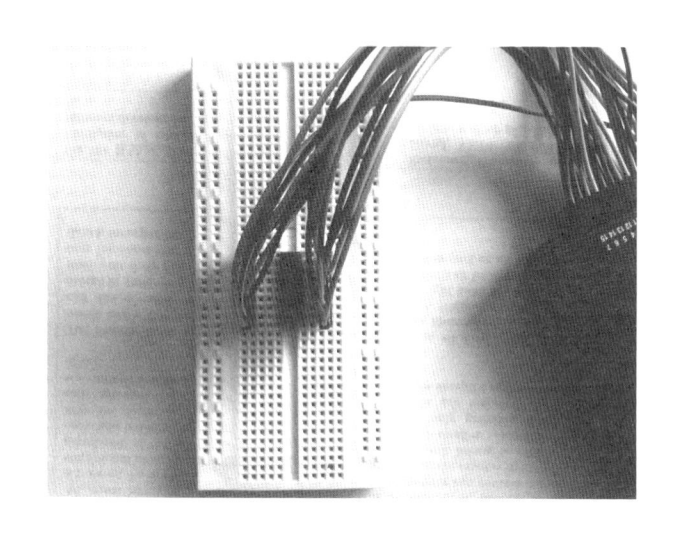

圖 5-9　　　將電路配置在不用銲接的麵包板上

圖 5-9 的 Analog Discovery Module，連同裝上了 WaveForms 軟體的 PC，成為多件由軟體操控的硬體儀器。測試數位電路，對於電路的輸入部份可用**數字模式產生器** (Digital Pattern Generator)，輸出部份可用**數位信號分析儀** (Digital Signal Analyzer)，有關 Digital Pattern Generator 和 Logic Analyzer 的設定，請參考附錄 C 和 D。圖 5-10 就是 WaveForms 軟體在 PC 上所顯示的選擇圖形。

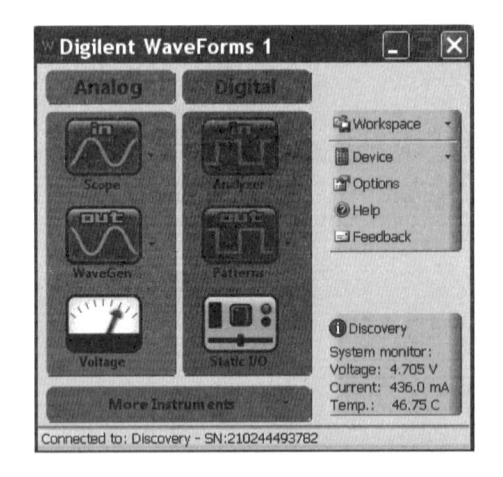

◎ 圖 5-10 ◎　　WaveForms 軟體在 PC 上所顯示的選擇圖形

點擊 Digital 部份的 out，便可獲得如圖 5-11 的數字模式產生器。

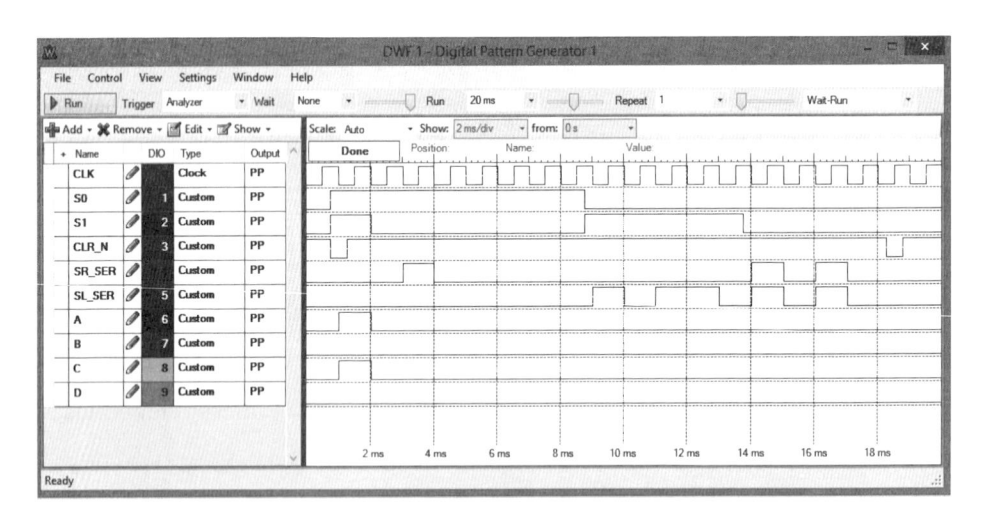

◎ 圖 5-11 ◎　　提供電路輸入訊號的數字模式產生器

有了測試 74LS194 電路的輸入訊號，還需要電路所需的 +5 V 直流電壓，點擊圖 5-10 WaveForms Analog 部份的 Voltage，可獲得圖 5-12 的二個 V+ 5 V 和 V– 5 V 直流電源，本實作只須 V+ ON。

◦ 圖 5-12 ◦　　　　WaveForms Analog 提供 +5 V 直流電壓

點擊 Digital 部份的 in，便可獲得如圖 5-13 的數位信號分析儀。

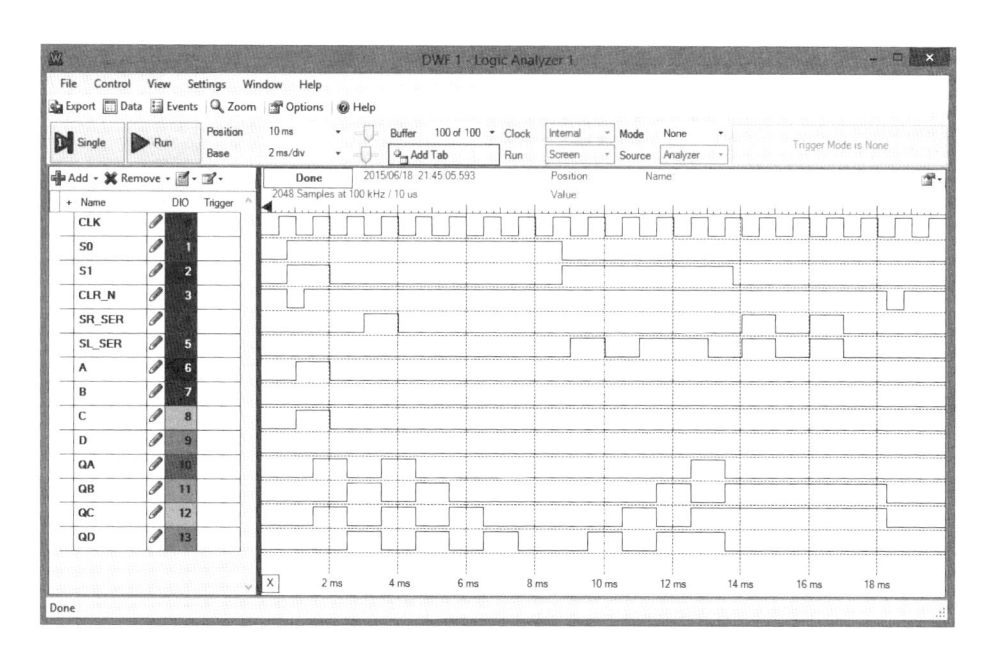

◦ 圖 5-13 ◦　　　　數位信號分析儀

5-5 課外練習

1. 試寫出 SN74LS194 的 VHDL Behavior Model 和測試它的 Testbench。並用 ModelSim Simulator 來測試證實之。

2. 試從圖 5-5 和圖 5-7 分析 74LS194 如何達成 Shift Right 和 Shift Left 的動作。請用簡化 Logic 說明之。

3. 試從圖 5-5 和圖 5-7 分析 74LS194 如何達成 Serial Input Shift Right 和 Serial Input Shift Left 的動作。請用簡化 Logic 說明之。

4. 試用 VHDL Behavior Model 寫出 SN74LS194 的模式。

5. 試用 VHDL 來寫出測試 SN74LS194 的 TestBench 模式，並用 ModelSim Simulator 來測試之。

第六章 計數器

計**數器** (Counter) 是數位系統中主要的子系統之一，由數位元件 T Flip Flop 所組成。系統中訊號的**計時器** (Timer)，就是靠計數器來完成，它是所有系統中不可缺少的組件。

6-1 DFF 組成的計數器

圖 6-1 的 Counter 是將 DFF 的 Q_N 連接到 D 所形成的 TFF 來組成的 4-bit UP Counter。

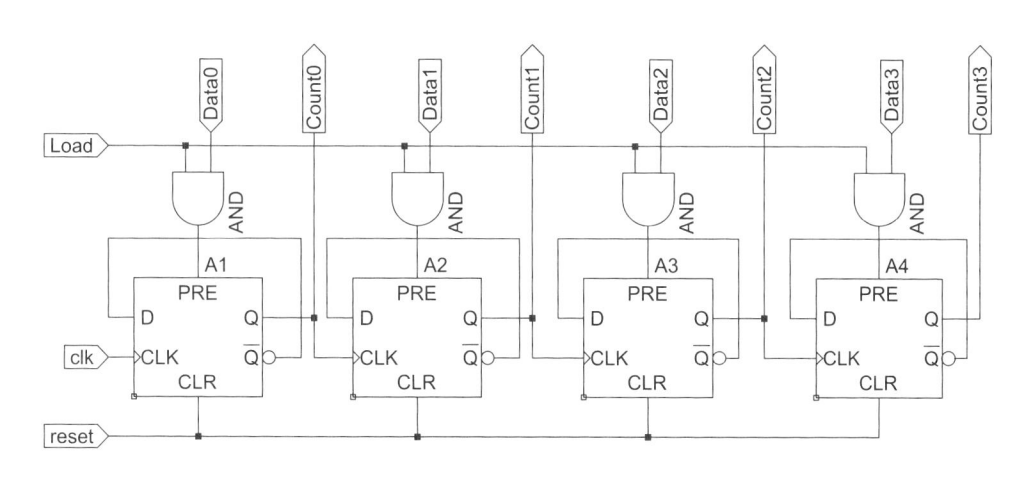

◦ 圖 6-1 ◦　　　　DFF 組成的 4-bit UP Counter

4-bit UP Counter 計數的執行，是由 0, 1, 2, 3……15, 0, 1, 2, 3… 如此周而復始。如果將 Q 連接到 CLK，改變成 Q_N 連接到 CLK 圖 6-1 的

55

Counter 就成為 Down Counter。4-bit Down Counter 計數的執行，是由 15, 14, 13, 12…….. 0, 15, 14, 13, 12… 數值由高而低地進行。JKFF 也可以將 J 和 K 都接 '1' 而成為 TFF 來組成 Counter。

◎ 6-2　使用 VHDL 來描述電路的結構和電路的模擬測試

對於較複雜的電路，除了使用硬體結構來描述電路，還可以用電路的 I/O **行為** (Behavior) 模式來對圖 6-1 做 VHDL 的電路的行為描述，如圖 6-2 所示。有關 VHDL 電路和 TestBench 的構成，請參考附錄 A 和 B。ModelSim 的 Simulation 步驟，請參考附錄 E。

```vhdl
-- 4bit UP counter -----------
library IEEE;
use IEEE.std_logic_1164.all,
    IEEE.numeric_std.all;
entity counter is
generic(n : NATURAL := 4);
port(clk : in std_logic;
     reset : in std_logic;
     load : in std_logic;
     Data: in unsigned(n-1 downto 0);
     count : out std_logic_vector(n-1 downto 0));
end entity counter;
architecture rtl of counter is
begin
    p0: process (clk, reset, load) is
    variable cnt : unsigned(n-1 downto 0);
    begin
        if reset = '1' then
            cnt := (others => '0');
        elsif load = '1' then
            cnt := Data;
        elsif rising_edge(clk) then
            cnt := cnt + 1;
        end if;
        count <= std_logic_vector(cnt);
    end process p0;
end architecture rtl;
```

圖 6-2　　4-bit UP Counter 的 VHDL Behavior 模式

模式中為了配合 Line 23，Counter 每接受一個 CLK 便加一的特性，所以 Line 10 Data 和 Line 16 cnt 要用 unsigned，而不能用 std_logic_vector。

圖 6-3 為 4-bit UP Counter 的 VHDL TestBench，首先將 10 加載到 4-bit Counter 上，由 10 向上計數。到 14 後被 reset 為 0，再重新計數。

```vhdl
1  --TstBench.vhd --------------------------
2  Library IEEE;
3  use IEEE.std_logic_1164.all;
4  IEEE.numeric_std.all;
5  entity TstBench is
6  generic(n : NATURAL := 4);
7  end TstBench;
8  ------------------------------------------
9  use work.all;
10 ------------------------------------------
11 architecture stimulus of TstBench is
12 --First, declare lower-level entity that to be test
13 component counter
14    port(clk : in std_logic;
15      reset : in std_logic;
16      load : in std_logic;
17      Data: in unsigned(n-1 downto 0);
18      count : out std_logic_vector(n-1 downto 0));
19 end component;
20 ------------------------------------------
21 --Next, declare TstBench's SINGNALs
22 SIGNAL clk: std_logic := '0';
23 SIGNAL reset: std_logic := '0';
24 SIGNAL  load: std_logic := '1';
25 SIGNAL  Data: unsigned(n-1 downto 0):= "1010";
26 SIGNAL count :  std_logic_vector(n-1 downto 0):= "0000";
27 ------------------------------------------
28 begin
29   DUT:  counter port map (clk, reset, load, Data, count);
30 ------------------------------------------
31 ---Concurrent Code for Periodical waveform
32   clk <= NOT clk AFTER 50 ns;
33   reset <= '1' AFTER 500 ns, '0' AFTER 600 ns;
34   load <= '0' AFTER 100 ns;
35 ------------------------------------------
36 end stimulus;
```

圖 6-3　　4-bit UP Counter 的 VHDL TestBench

測試的結果如圖 6-4 所示。全部測試時間為 2.5 ms，4-bit Counter 完成由 0 至 15 計數然後自動歸 0 後再 Up Count。

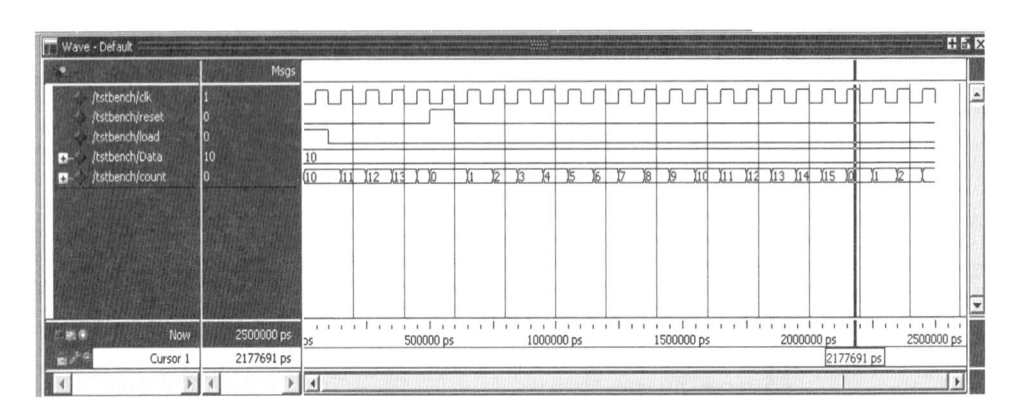

圖 6-4　4-bit UP Counter 測試的結果

6-3　74LS193 Synchronous 4-Bit Binary Counter with Dual Clock 簡介

從 Counter 的種類和功用來講 74LS193 是一個完善的 4-bit Counter，它不但具有同步獨立的 Up Count 和 Down Count，同時能夠 Load Data、reset Data 等功能。圖 6-5 是它的電路結構。4-bit Counter 部份，主要使用 4 個 TFF。其他如同步、Up Count、Down Count、Load、Clear 是用 Gates 的 Concurrent 特性來完成。

圖 6-5 為 74LS193 的電路結構圖。圖 6-6 是電路的輸入與輸出的邏輯時間 Timing 關係圖。在下一節硬體實作裡，就將依據這個 Timing 關係來設定 Logic Pattern Generator 的輸出波形。

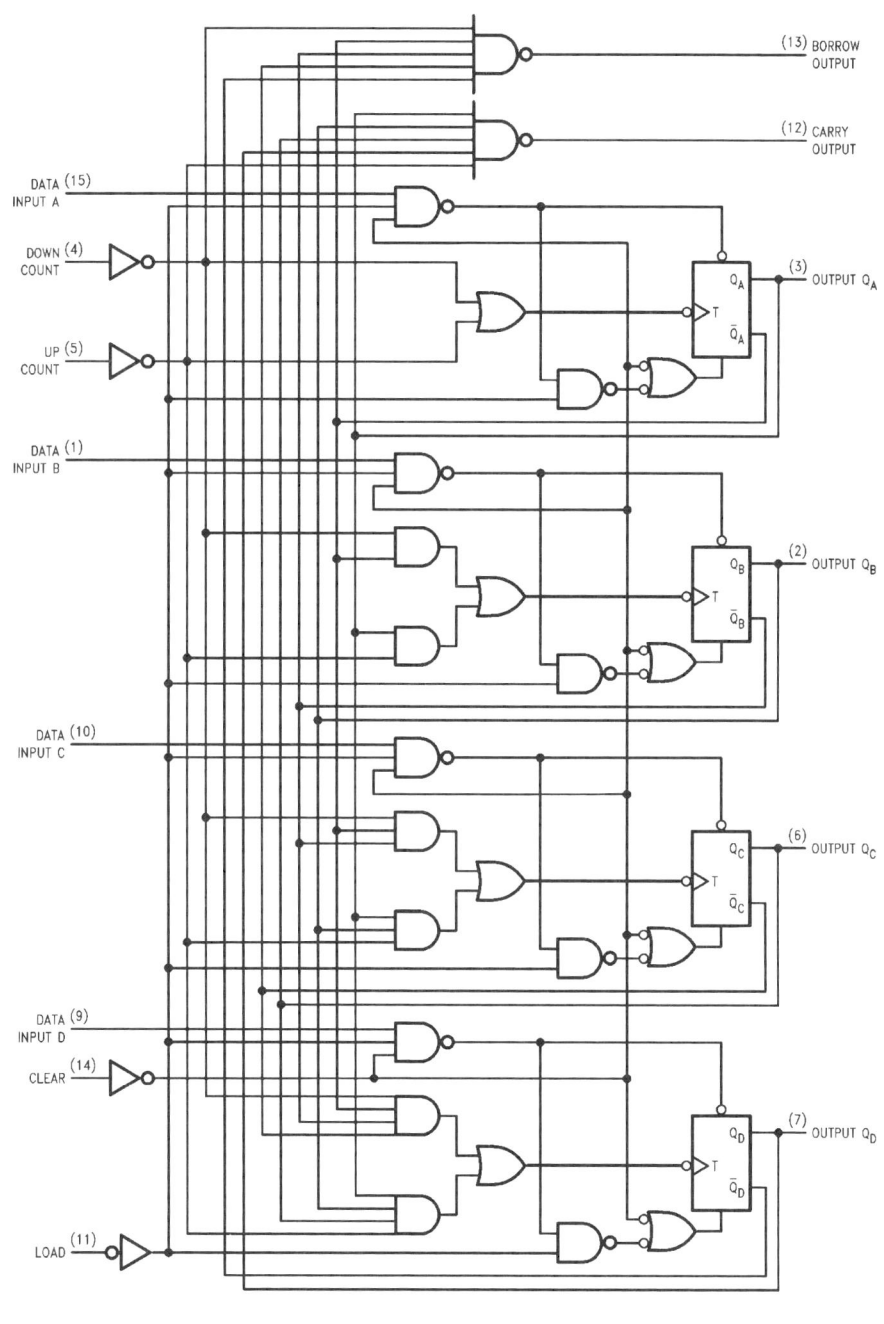

圖 6-5　74LS193 的電路結構

Timing Diagram

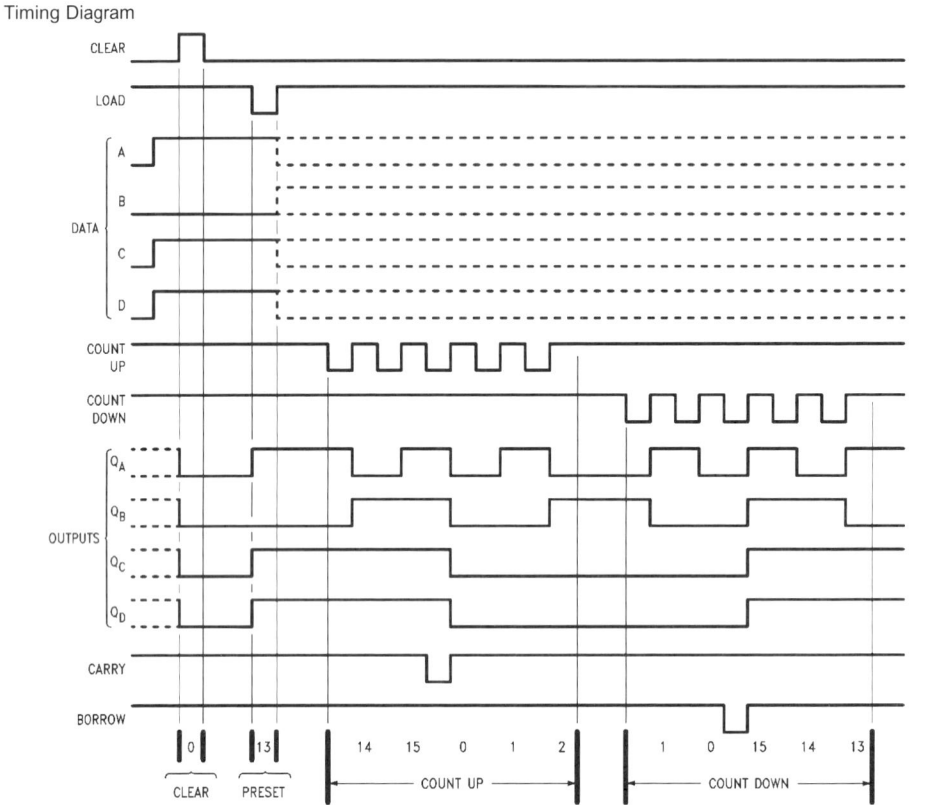

Note A: Clear overrides load, data, and count inputs
Note B: When counting up, count-down input must be HIGH; when counting down, count-up input must be HIGH.

圖 6-6　電路的輸入與輸出的邏輯時間 Timing 關係圖

6-4　硬體實作

圖 6-7 為 Analog Discovery Module 與 SN74LS193 的測試連接的電路圖。它的輸入為 Clear、Load_N、Data Ain ~ Data Din、Count Up、Count Down 等 8 個訊號。輸出為 QA ~ QD、Carry_N、Borrow_N 等 6 個訊號。全部共 14 個訊號。這一個實作是做 6.3 節測試 74LS193 Counter 的 Clear、Set Data、Count Up 和 Count Down 等功能。

第六章 計數器

```
Analog Discovery to 74LS193
        connections
RED    →  pin 16  Vcc
DIO 0  →  pin 14  Clear
DIO 1  →  pin 11  Load_N
DIO 2  →  pin 15  Data Ain
DIO 3  →  pin  1  Data Bin
DIO 4  →  pin 10  Data Cin
DIO 5  →  pin  9  Data Din
DIO 6  →  pin  5  Count UP
DIO 7  →  pin  4  Count Down
DIO 8  →  pin  3  QA
DIO 9  →  pin  2  QB
DIO 10 →  pin  6  QC
DIO 11 →  pin  7  QD
DIO 12 →  pin 12  Carry_N
DIO 13 →  pin 13  Borrow_N
BLACK  →  pin  8  GND
```

```
                U1
  1 Data Bin      V_CC    16
  2 QB         Data Ain   15
  3 QA           Clear    14
  4 Count Down            13
               Borrow_N
  5 Count UP              12
                Carry_N
  6 QC           Load_N   11
  7 QD         Data Cin   10
  8 GND        Data Din    9
```

圖 6-7　SN74LS193 功能測試與 Analog Discovery 連接圖

　　簡單的數位電路，如果使用比較低的頻率 (1 KHz)，可以使用不用銲接的**麵包板** (Breadboard) 來完成，如圖 6-8 所示。頻率較高 (100 KHz) 以上或比較複雜的電路，應當把電路裝配在接地良好、使用銲接的電路板上。

圖 6-8　將電路配置在不用銲接的麵包板上

使用 Analog Discovery Module，連同裝上了 WaveForms 軟體的 PC，成為多件由軟體操控的硬體儀器。測試數位電路，對於電路的輸入部份可用**數字模式產生器** (Digital Pattern Generator)，輸出部份可用**數位信號分析儀** (Digital Logic Analyzer)。有關 Digital Pattern Generator 和 Logic Analyzer 的設定，請參考附錄 C 和 D。圖 6-9 就是 WaveForms 軟體在 PC 上所顯示的選擇圖形。

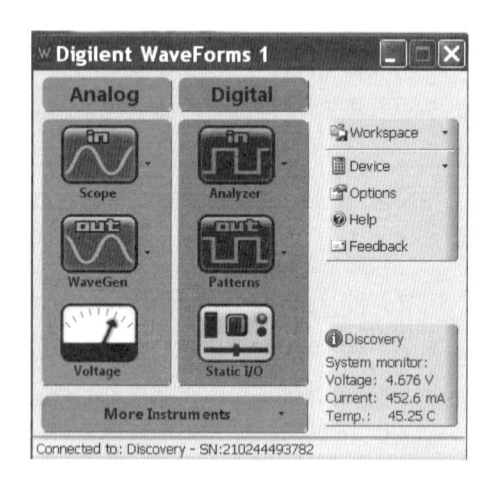

圖 6-9　WaveForms 軟體在 PC 上所顯示的選擇圖形

點擊 Digital 部份的 out，便可獲得如圖 6-10 的數字模式產生器。數字模式產生器的設定如下：

圖 6-7 為 SN74LS193 與 Analog Discovery Module 間的連接圖。它的輸入依次為 CLEAR、LOAD、DATA A, B, C, D 和 COUNT UP、COUNT DOWN。輸出為 QA、QB、QC、QD 和 CARRY 及 BORROW。信號的排列次序如圖 6-10。

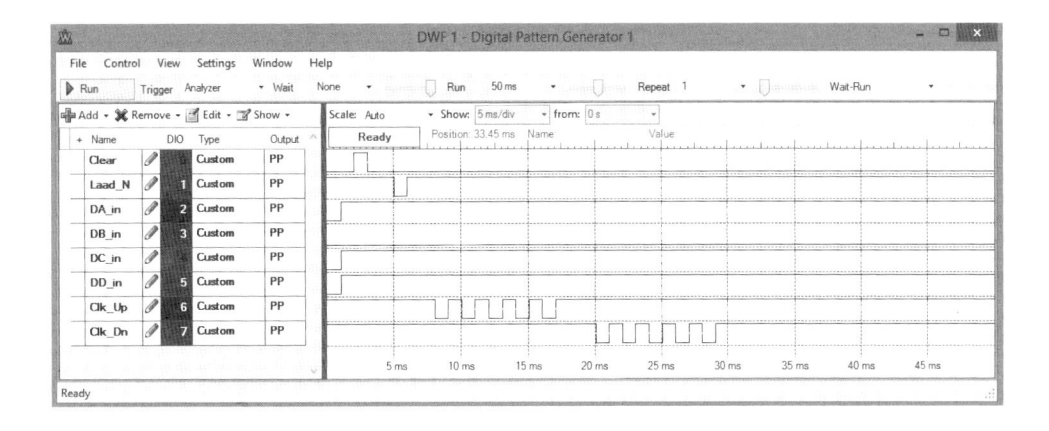

圖 6-10　　Pattern Generator 的信號排列次序

　　74LS193 的輸出訊號依次為 QA, QB, QC, 和 QD 加上 Carry 及 Borrow，為了能夠讀出 Count Up 和 Count Down 的數字，選用 **BUS Q/Decimal**。

　　點擊 **Digital** 部份的 **in**，便可獲得數位信號分析儀。為了容易鑑別 I/O 訊號間的 Timing 關係。可以把 **Pattern Generator** 的輸入訊號也合併在內，結果使用了 DIO 0～DIO 13 等 14 個訊號，如圖 6-11 所示。

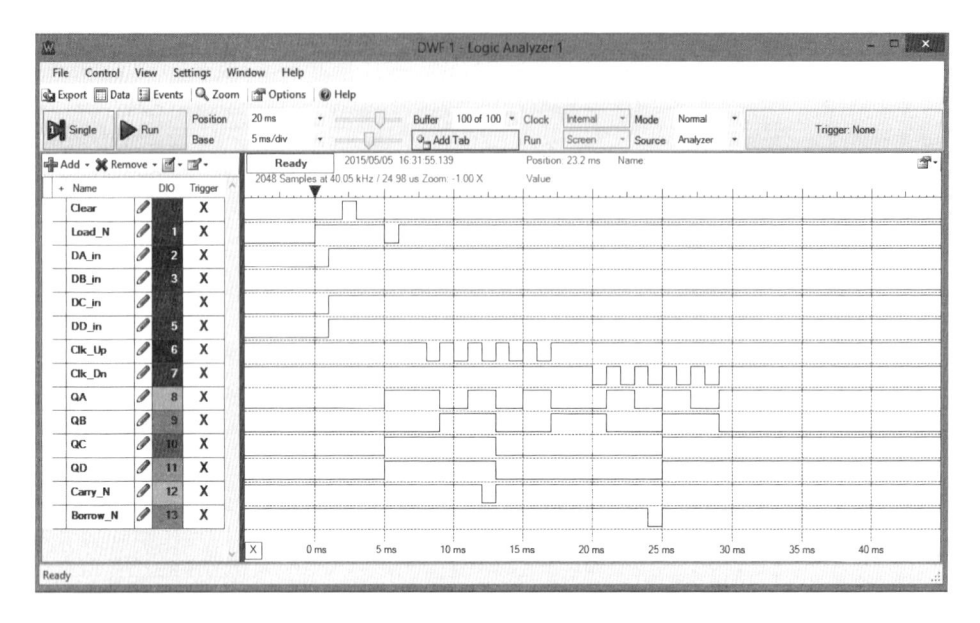

64　iLAB Digital 數位電路設計、模擬測試與硬體除錯

◦ 圖 6-11 ◦　　Logic Analyzer 顯示測試 74LS193 的全部 I/O 訊號

執行 **Logic Analyzer** 的測試工作，除了要準備以上的安排及設定之外，還有 74LS193 TTL 所需的 V+ 5V 電源。點擊圖 6-9 **WaveForm** 的 **Voltage** 當可獲得電源如圖 6-12 所示。選用 V+ ON 及 Power ON。

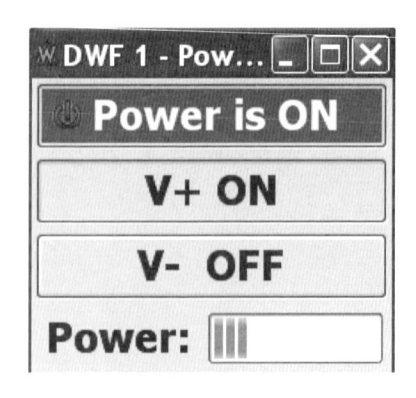

◦ 圖 6-12 ◦　　TTL 74LS193 只需用 V+ 5V 電源

6-5 課外練習

1. 74LS193 是一個 Binary Counter，計數由 0～15 或 15～0 進行。試改變它的計數由 0～9 或 9～0 進行。(可以外加 Gate)

2. 試將圖 6-2 4-bit Binary UP Counter 的 VHDL Behavior 模式，改寫成 Decade Counter 的 VHDL 模式，並用 ModelSim Simulator 來測試之。

3. TTL 的 7 segment display IC 7447 常用來配合 Counter 顯示讀數，試寫出其 4-bit Binary Code 轉換成 Hex Code 的 VHDL Behavior 模式。

4. 試將圖 6-2 4-bit UP Counter 的 VHDL Behavior 模式，改寫成 Down Counter 的模式，並用 ModelSim Simulator 來測試之。

5. 試用線路圖來表明如何可將二進制計數器 SN74LS193 連接成十進制計數器。(提示：可以外加各式 Gate)

6. 試用 VHDL 設計一計數器可以從 8 MHz 的 System Clock 得到 1 Hz 的 Clock。

7. 試用 Block diagram 來表示一個 10 MHz Frequency Counter 的構成。

第七章　數位訊號轉換成類比訊號

大自然中所有"熱聲光電"無一不是以"類比訊號"的形式存在。"數位訊號"只是近年來，人們在處理自然界許多事物時的一種方法。它的開始與終了，還得回歸到"類比訊號"。

7-1　階梯式 R2R 電阻組成的 D to A 轉換器

D to A，是數位進、類比出的轉換器，種類很多。最簡單的當屬由電阻組成的 R2R 階梯式 D to A 轉換器，如圖 7-1 的電路所示。

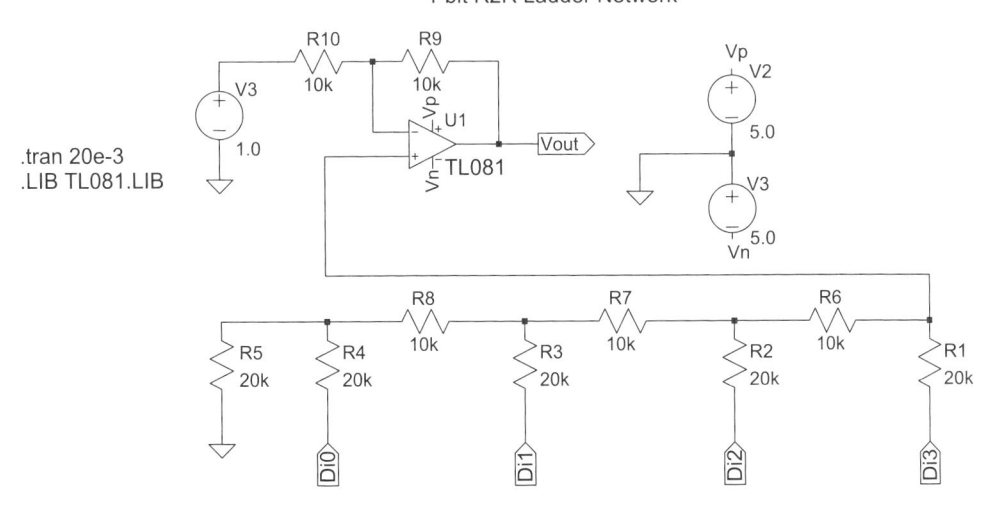

圖 7-1　電阻組成的 4-bit R2R 階梯式 D to A 轉換器電路

這個 4-bit 的 D to A，它的數位輸入是由 Di0~Di3 所組成。Vref 用來控制類比輸出電壓的大小。如果 Vref = 1.0 V；由於 LTspice 中的邏輯電路 Di0~Di3 的 '1' 為 1.0 V；'0' 為 0 V；那麼 Vout 的輸出範圍為 −1.0 V ~ +1.0 V。倘若把 Vref 和 R18 移除；那麼 Vout 的輸出範圍為 0 V ~ +1.0 V。圖 7-2 R2R 階梯式 D to A 轉換器電路的測試，是在 LTspice Simulator 下進行。其中 Vref = 1.0 V；Vout 的輸出範圍為 −1.0 V ~ +1.0 V。當 D0~D3 = "0001" 時，圖 7-3 為測試的結果，當為 0 mV。

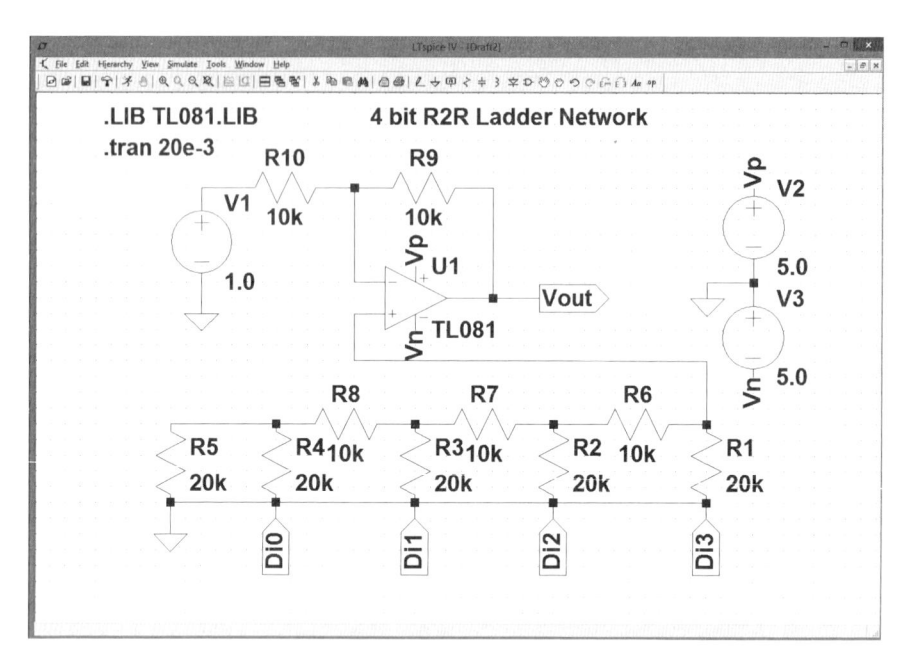

圖 7-2　R2R D0 ~ D3 = "0001" 轉換器電路的測試

第七章　數位訊號轉換成類比訊號　69

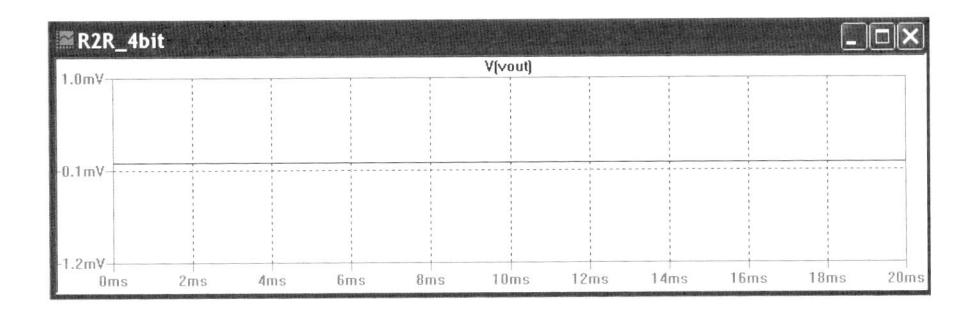

◦ 圖 7-3 ◦　　　測試的結果　Vout = 0 mV

　　測試 R2R 4-bit D to A 最好的方法是把 Di0～Di3 連接到一個 4-bit **計數器** (counter) 的輸出 D0～D3 上來進行。或者是利用 Analog Discovery Module 的 Digital Pattern Generator 的 4-bit Binary Counter 所產生的波形來完成。

　　LTspice 主要是一個 Analog Simulator，也有部份的 Logic 元件，可供混合 Analog + Digital (Mixed) 電路 Simulation 之用。圖 7-4 就是一個例子，它把圖 7-2 R2R D0～D3 加上由 4 個 DFF 連接成 TFF 組成的 UP Counter，Q0～Q3 經由 Buffer A5～A8 接到 D0～D3 來做測試。

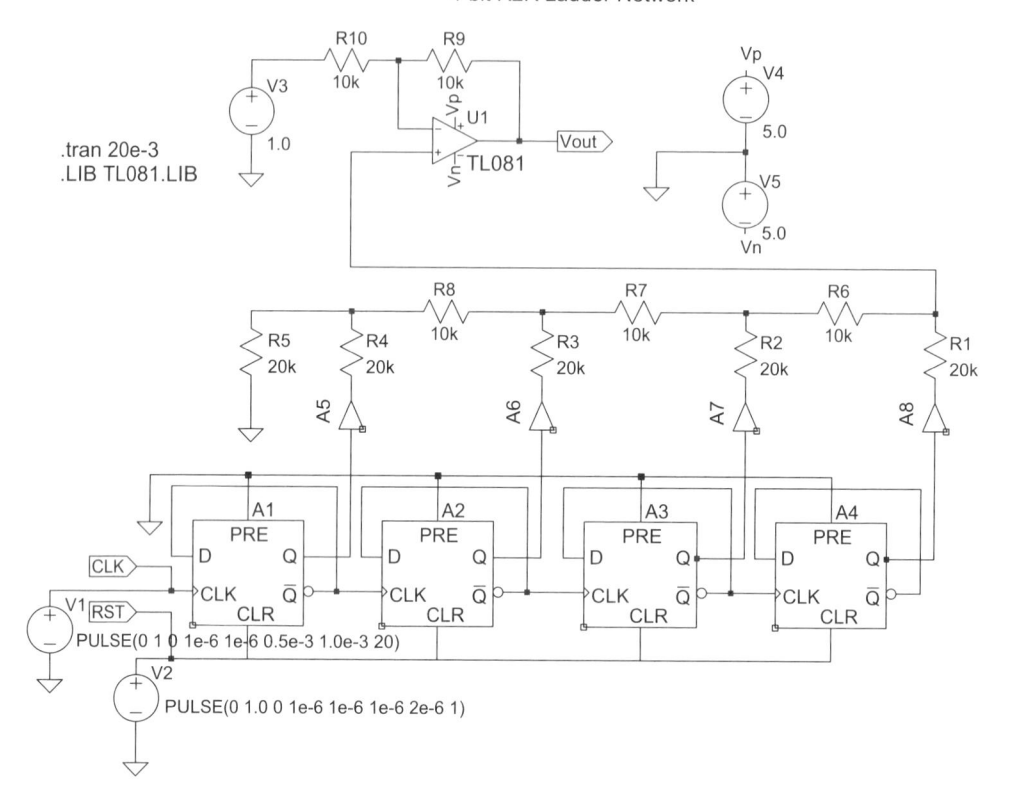

圖 7-4　4-bit UP Counter 提供測試訊號給 4-bit R2R D to A 轉換電路

測試的結果如圖 7-5 所示。清楚地顯示出由 4-bit 的 "0000" 到 "1111" 造成的 16 梯階。

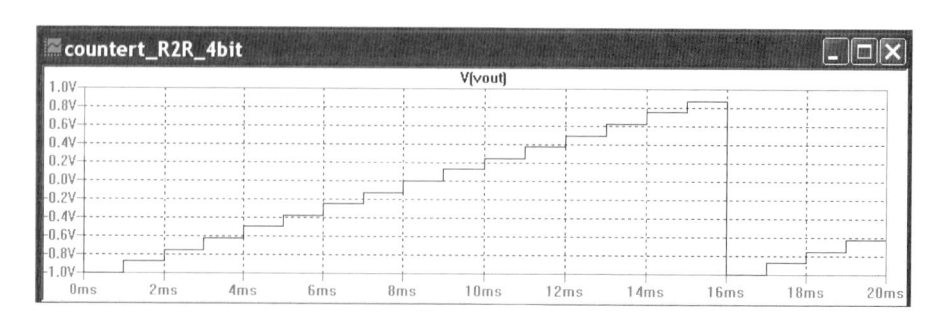

圖 7-5　4-bit R2R D to A 轉換電路測試的結果

7-2　D to A 積體電路 DAC0808 的介紹

　　積體電路的 8 bit D to A 轉換器，由於無法製出 R2R 電路所需的精確電阻，設計上改用電流開關來達到相同的效果。

　　圖 7-6 為 DAC0808 結構示意圖，圖 7-7 為 DAC0808 積體電路。清晰的圖片請參考下載 DAC0808 的規格特性[1]。

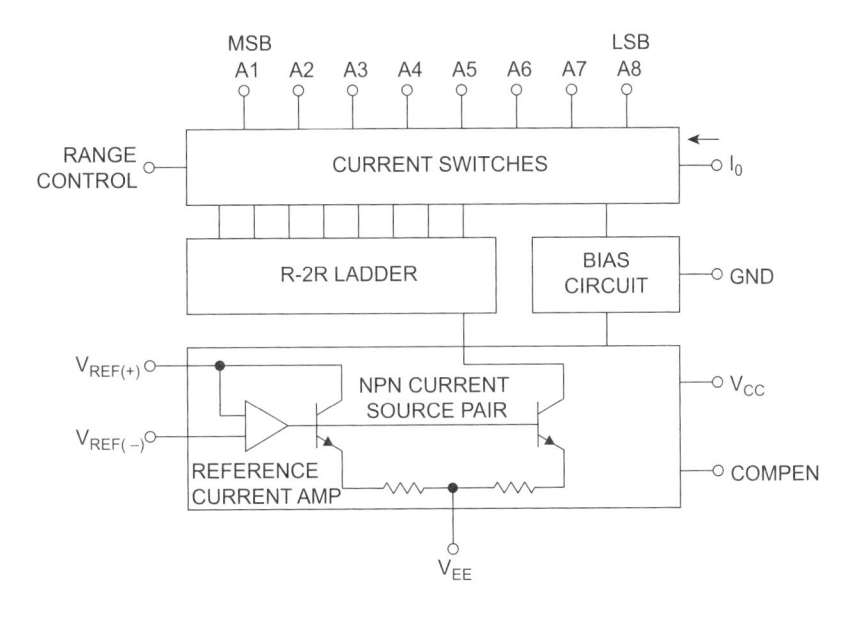

圖 7-6　DAC0808 結構示意圖

---

[1] http://www.analog.com/static/imported-files/application_notes/80635169AN17.pdf

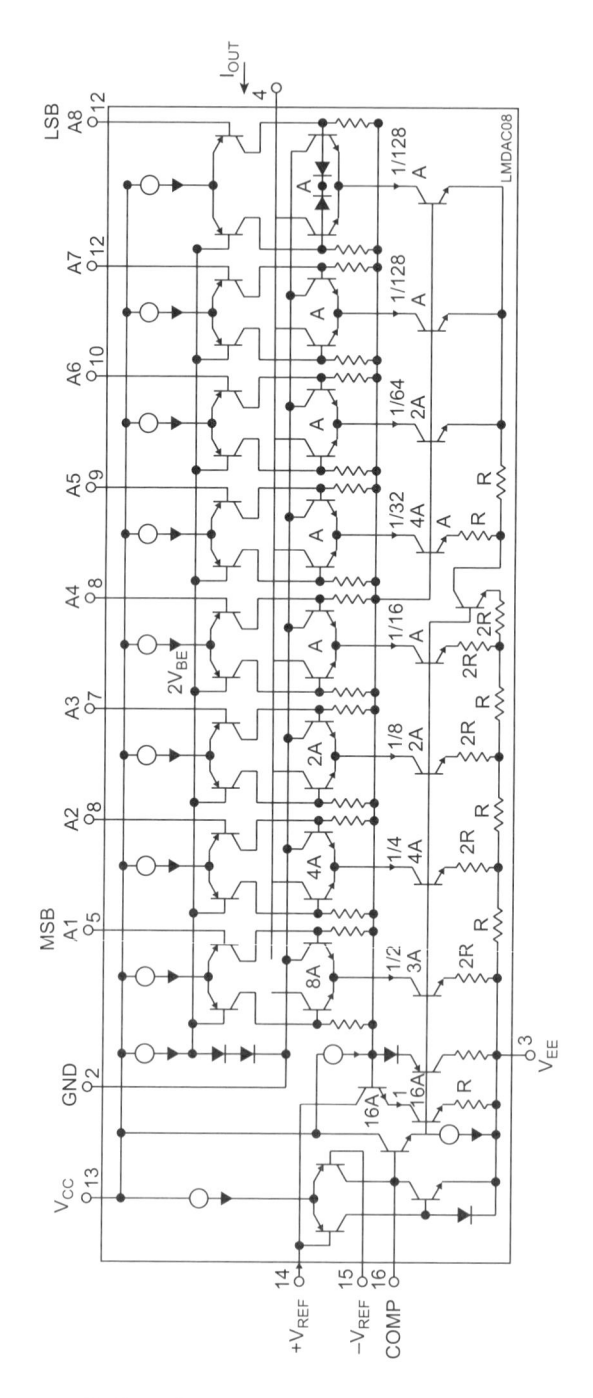

圖 7-7　DAC0808 積體電路

圖 7-8 為 DAC0808 積體電路的測試裝置，數位訊號 '1' 或 '0' 加到 A1~A8 接腳上，輸出電流 I0 和 數位訊號輸入的關係為：

$$I0 = K\left(\frac{A1}{2} + \frac{A2}{4} + \frac{A3}{8} + \frac{A4}{16} + \frac{A5}{32} + \frac{A6}{64} + \frac{A7}{128} + \frac{A8}{256}\right)$$

其中 k = Vref / R14，電路圖中 R15 作為溫度補償之用。

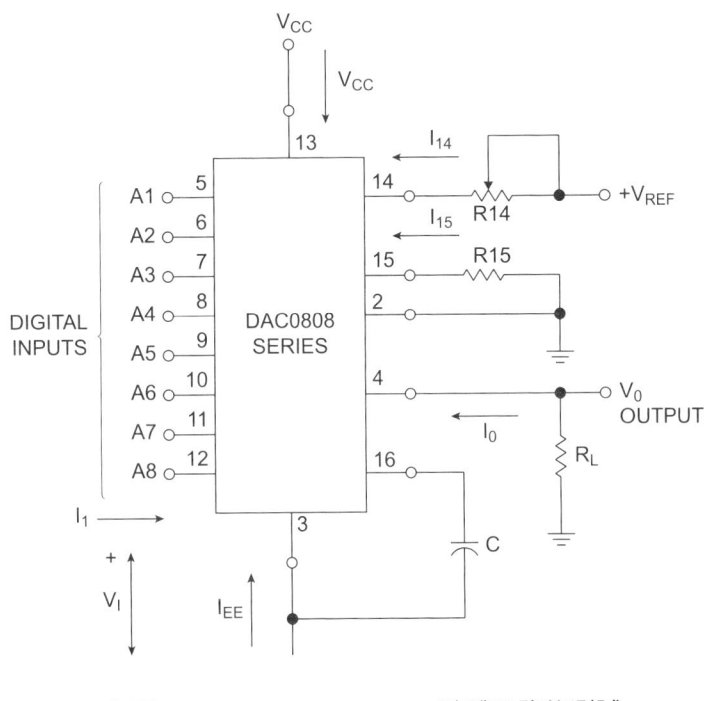

圖 7-8　DAC0808 積體電路的測試

7-3 硬體實作

圖 7-9 為 Analog Discovery Module 與 DAC0808 的測試連接的電路。由於是 DAC，所以須用到 Module 中的 Digital 和 Analog 二方面的

測試儀器。Analog Discovery Module 中的 Digital 提供 DAC0808 輸入所需的 8-bit Counter D7~D0 訊號。電路的 Vcc 和 Vee 由 Module 的電源 +5 V 和 –5 V 提供。DAC0808 的輸出電流 –I0，經 TL081 運算放大器轉變成電壓之後由 TL081 的 Pin 6 輸出，實作的安排是用示波器 2+ 來測試。

圖 7-4 使用 4-bit UP Counter 提供測試訊號給 4-bit R2R D to A 轉換電路，DAC0808 的輸入為 8 bit，由於 Analog Discovery 的 Digital Pattern Generator 可以提供多至 16-bit Counter 的輸出，所以可以直接用它來測試 DAC0808。圖 7-9 左邊便是 Analog Discovery 的測試連線關係。

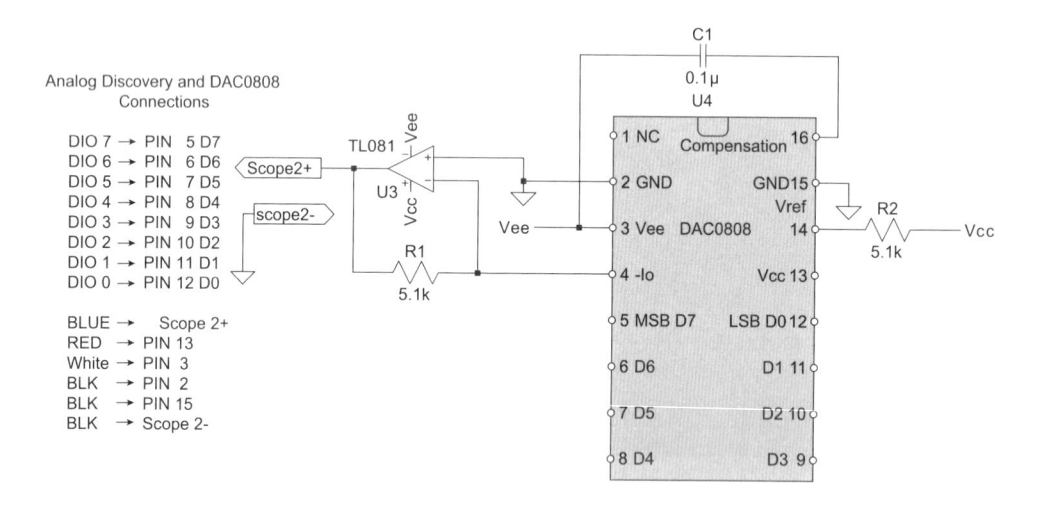

◎ 圖 7-9　　Analog Discovery Module 與 DAC0808 的測試連接

圖 7-10 為它們在麵包板的零件佈置。Analog Discovery Module 可以直接連接到 IC 二旁的接線柱上，如圖 7-11 所示。

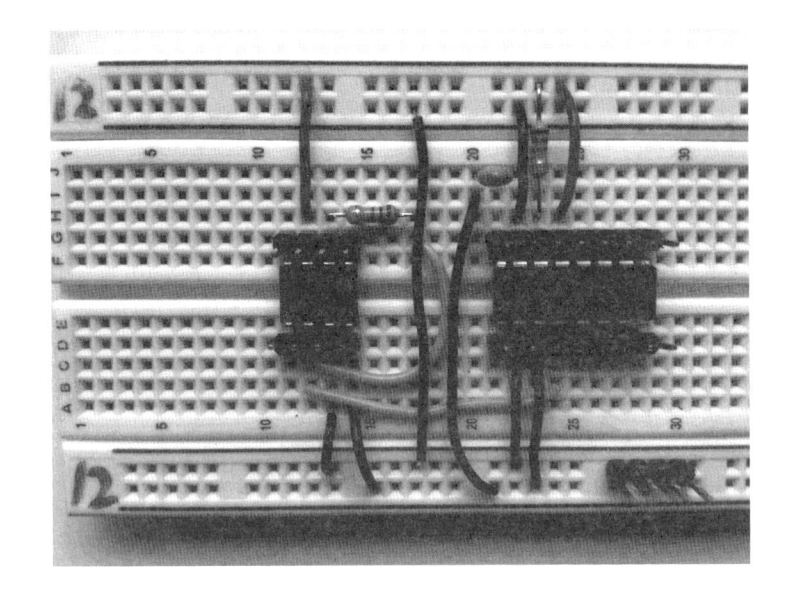

圖 7-10　電路在麵包板的零件佈置

圖 7-11　Analog Discovery Module 的接線直接連接到 IC 二旁的接線柱上

DAC0808 的測試，除了如圖 7-12 必要的電源之外，其它只需要圖 7-17 的 Digital Pattern Generator 和圖 7-18 的示波器。

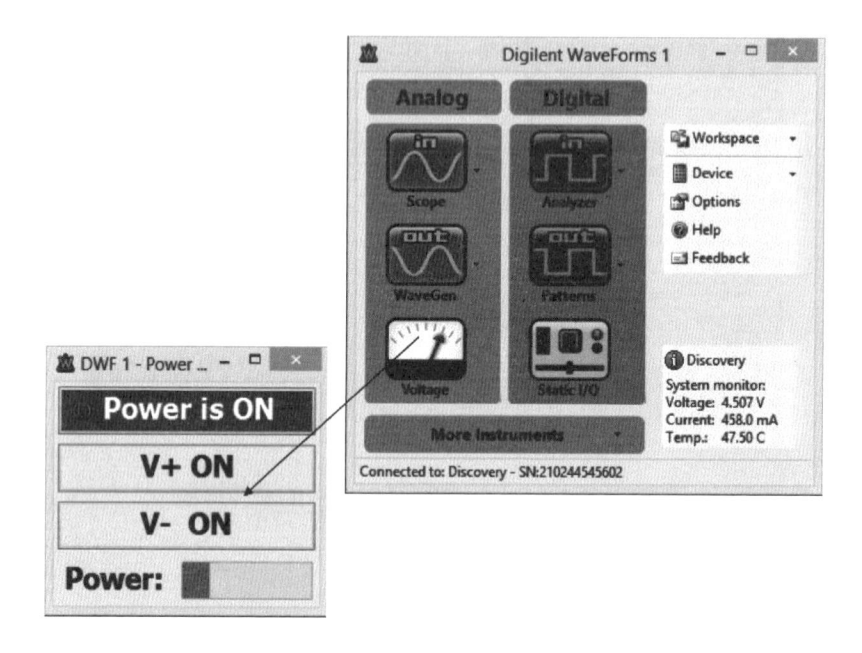

◦◦ 圖 7-12 ◦◦　　　測試所需之電源

測試 DAC0808 需要先設定 Pattern Generator 使其產生 8-bit、1 KHz 的 Binary Counter 的輸出。步驟如下：

1. 首先在圖 7-12 的 Digital waveform 上點擊 **Digital** 的 **OUT**，得圖 7-13 之 Pattern Generator 空白圖。

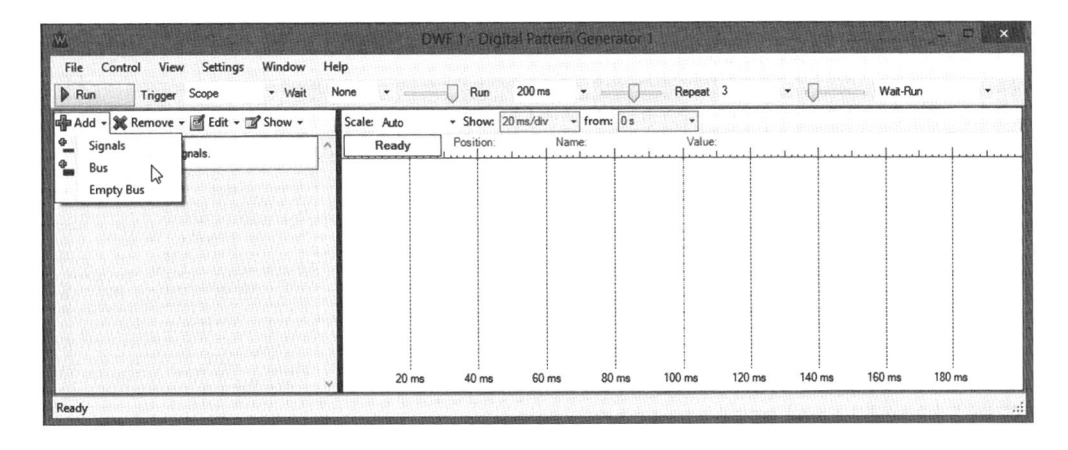

第七章　數位訊號轉換成類比訊號　77

❄圖 7-13❄　　　空白之 Pattern Generator 圖

2. 再點擊圖 7-13 之 +Add，並選用其中之 Bus，得圖 7-14 之 Add Bus 特性表。其中 **Format** 應選 **Binary**。LSB 為 DIO 0。

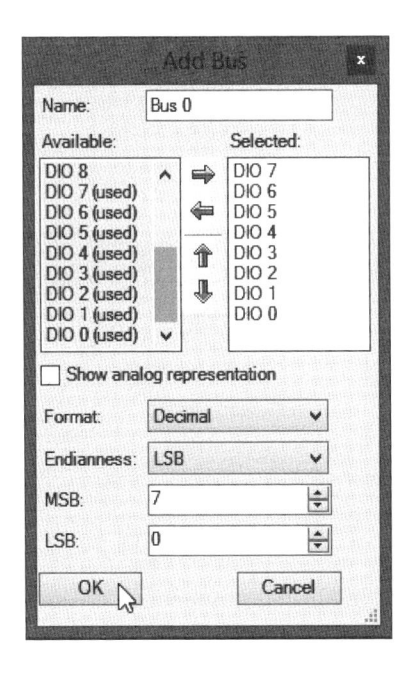

❄圖 7-14❄　　　Bus 的特性表

3. 再點擊 OK，得圖 7-15 之 DW1 Pattern Generator 1 空白圖。

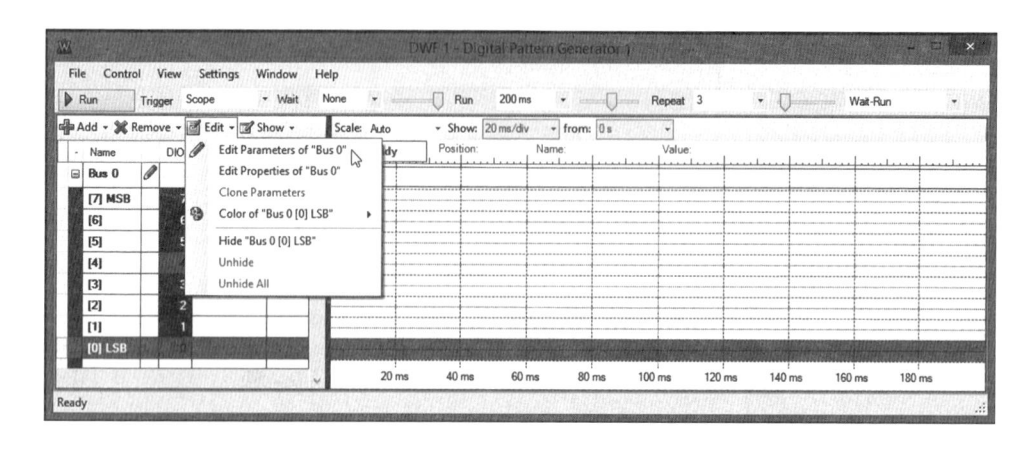

◦ 圖 7-15 ◦　　DW1 Pattern Generator 1 空白圖。

4. High Light Bus 的 DIO 0，並選用圖 7-15 的 Edit parameter of Bus 0 得圖 7-16 之 Bus 0 規格表。

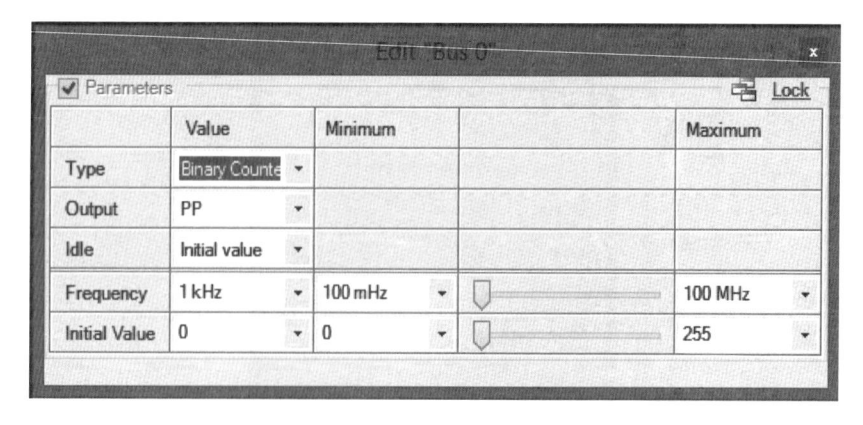

◦ 圖 7-16 ◦　　Bus 0 的規格表

在 Bus 0 的規格表中 Type 項選用 **Binary Counter**，Frequency 為 1 KHz，Idle 為 Low，Output 為 PP。結果將獲得測試 DAC0808 的 8-bit Binary Counter 的 Digital Pattern Output，如圖 7-17 所示。

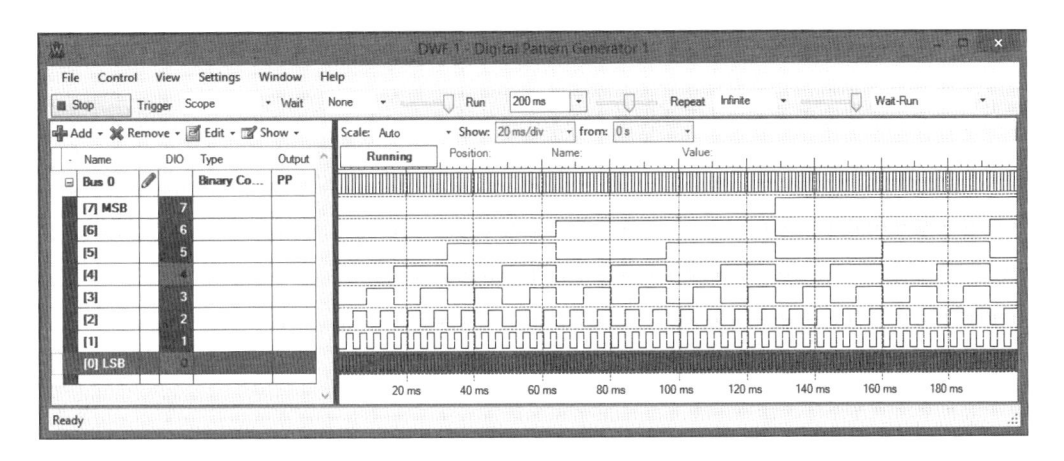

◦ 圖 7-17 ◦　　測試 DAC 0808 的 8-bit Binary Counter 的 Digital Pattern

DAC0808 在 8 bit Binary Counter 的輸入情況下，再經由 TL081 的電流到電壓的轉換，當獲得如圖 7-18 的波形。

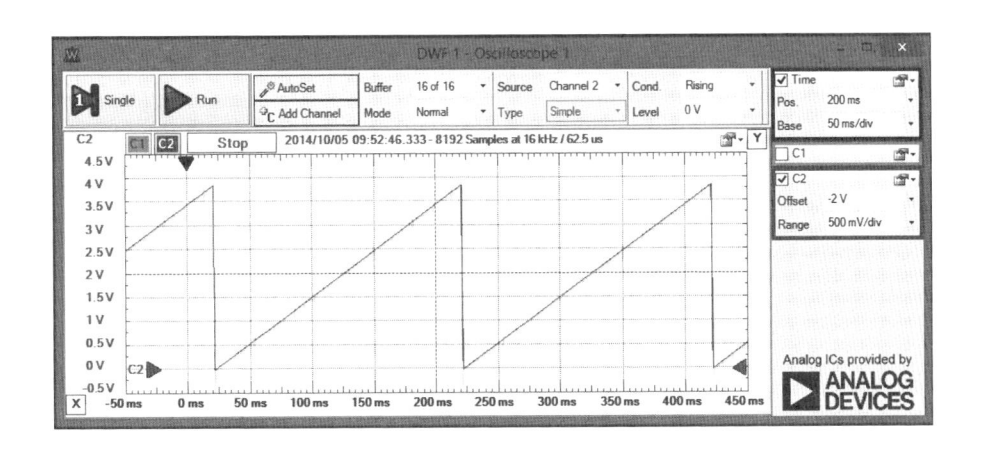

◦ 圖 7-18 ◦　　用示波器 Channel 2 以觀察測試 DAC0808 的結果

7-4 課外練習

1. 試參考 [1] Analog Devices AN-17 Application Notes，敘述如何使用 DAC-08 來完成 8x8 multiplication of two digital words。

2. 試述如何使用 DAC-08 來完成 Sine Waveform Generator。

3. 試製作一 LooK Up Table 以便 DAC0808 Sine Waveform Generator 的輸入所需。

4. 試參考 [1] Analog Devices AN-17 Application Notes p.8 Figure 27，說明使用 DAC-08 和 AM2505 來組成 A/D Converter 的原理。

第八章　類比訊號轉換成數位訊號

　　類比訊號轉換成數位訊號的方法，有 Flash/parallel，Ramp/counting 和 Successive approximation 等多種。它們都較數位訊號轉換成類比訊號的電路結構來得複雜。由於數位輸出的鑑別率，與 bits 的多寡有關，數位取樣的頻率與被取樣的訊號頻率有關。使用時對於 A to D 的特性和限制，必須事先瞭解[1]。

◎ 8-1　ADC0804 的介紹

　　ADC0804 結構上屬於 SAR (Successive Approximation Register)，它的基本原理如圖 8-1 所示[2]。

---

[1]　http://194.81.104.27/~brian/DSP/ADC_notes.pdf

[2]　http://www.analog.com/library/analogdialogue/archives/47-12/valid_first_conversion.html

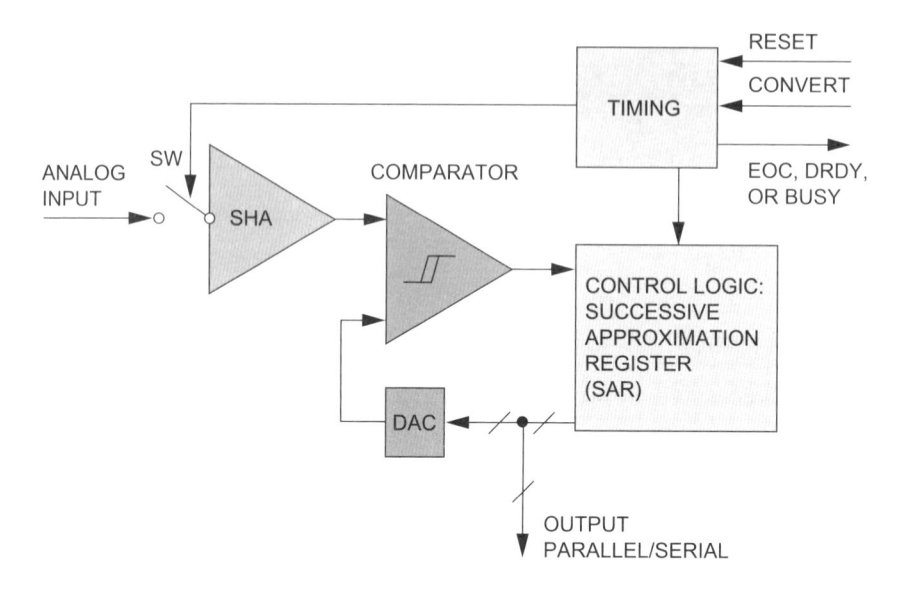

圖 8-1　SAR A2D 的基本原理

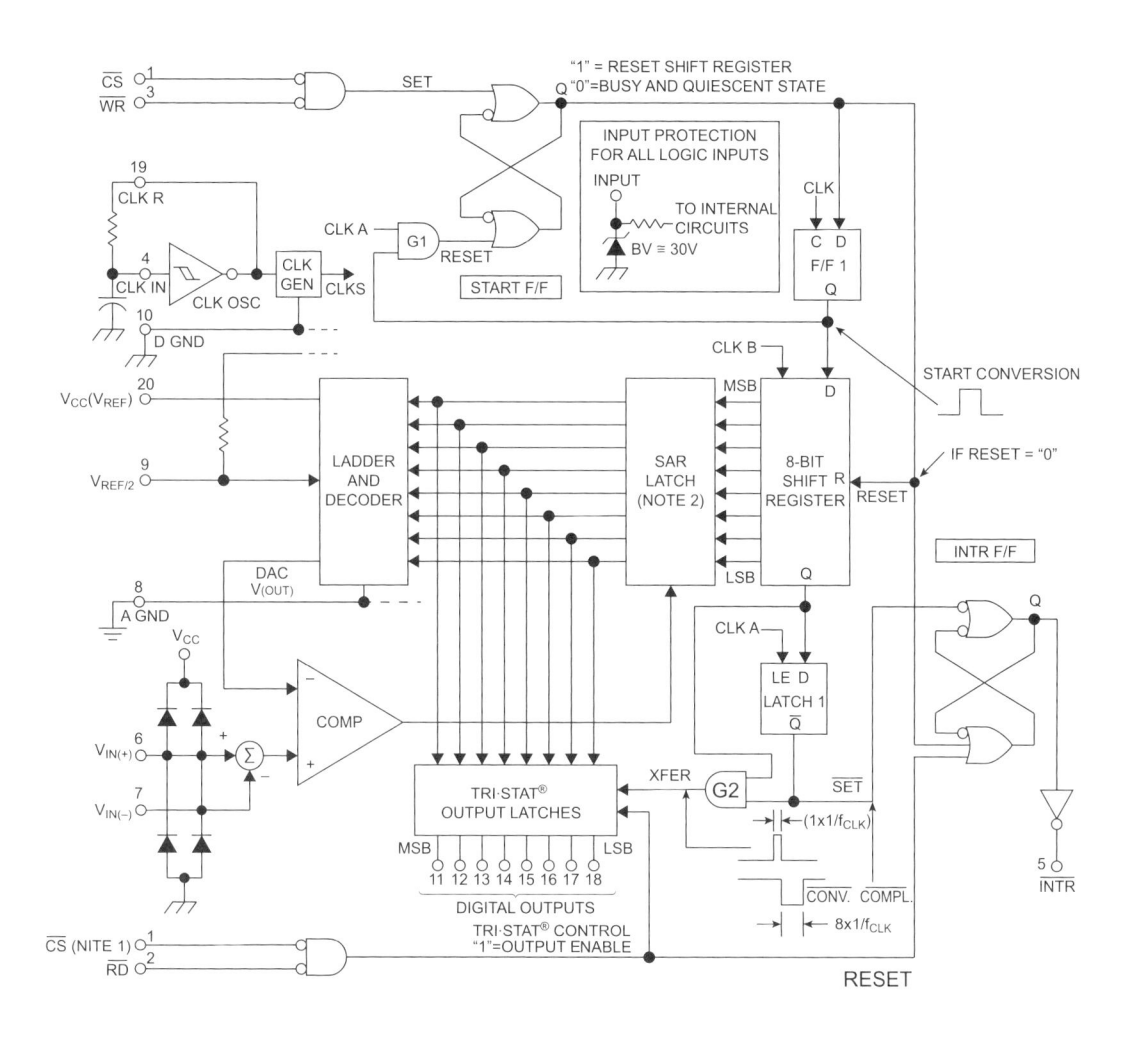

第八章　類比訊號轉換成數位訊號　83

圖 8-2　　ADC0804 的詳細結構方塊圖

　　這一類的 A2D IC 屬於中階，一般速度在 5 Msps 左右，解析度可達 16 bits，是目前 A2D 中售價較低的一種。多與微控制器合併使用。

8-2 ADC0804 的測試

ADC0804 的測試最簡單的方法[3]，是將一個已知的電壓接到 ADC0804 的輸入端，用目測來判別它的 8 個 LED 數碼的錯對。如圖 8-3 所示。為了易於測試，$V_{REF}/2$ (pin 9) 的輸入電壓應為 2.560 V，Vcc 為 5.12 V。則輸出的 LSB bit 相當於 20 mV。

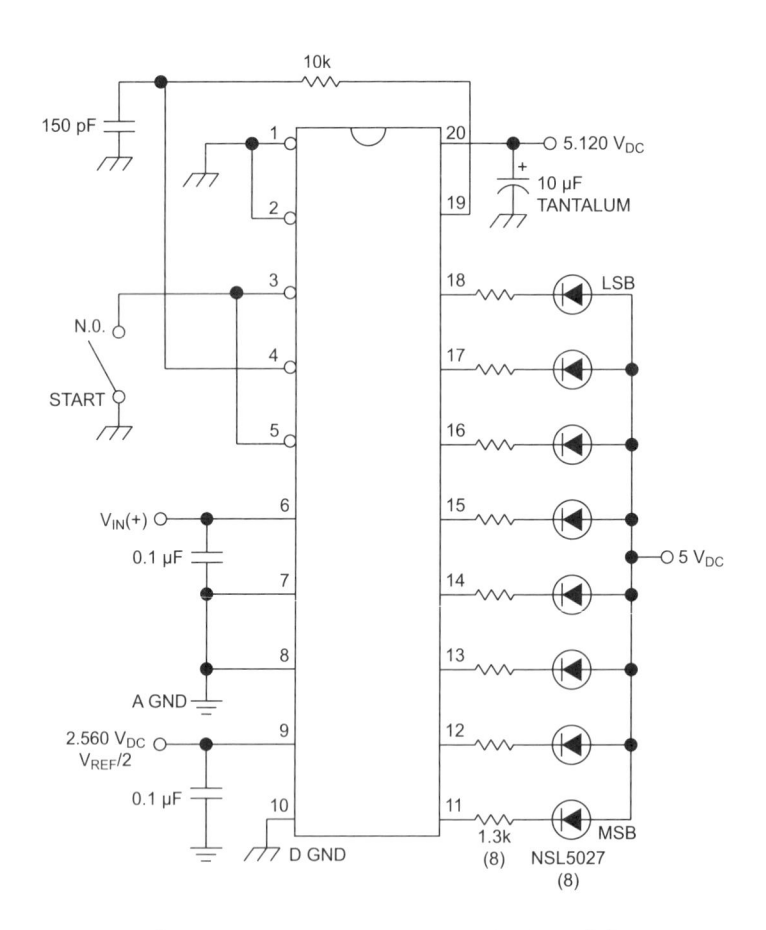

圖 8-3　ADC0804 的簡易測試

---

[3] http://www.alldatasheet.com/datasheet-pdf/pdf/8105/NSC/ADC0804.html

A2D ADC0804 如果結合第七章的 D2A DAC0808，除了如圖 8-3 的靜態測試之外，還可以做系統性的動態測試，當在硬體實作時再實施。

8-3 硬體實作

圖 8-4 為 ADC0804–DAC0808 的聯合測試電路。其中 U1 為 D2A，U2 為 A2D，U3 為輸出運算放大器，U4 為輸入運算放大器，輸入電壓經 U4 緩衝後來到 U2 A2D 的輸入端，U2 的數位輸出再接到 U1 的輸入，D2A 的 Analog 電流輸出再經由 U3 的電流轉變成電壓，最後從 pin6 輸出到示波器 2+。實作的目標是輸出的波形跟輸入的波形相似。

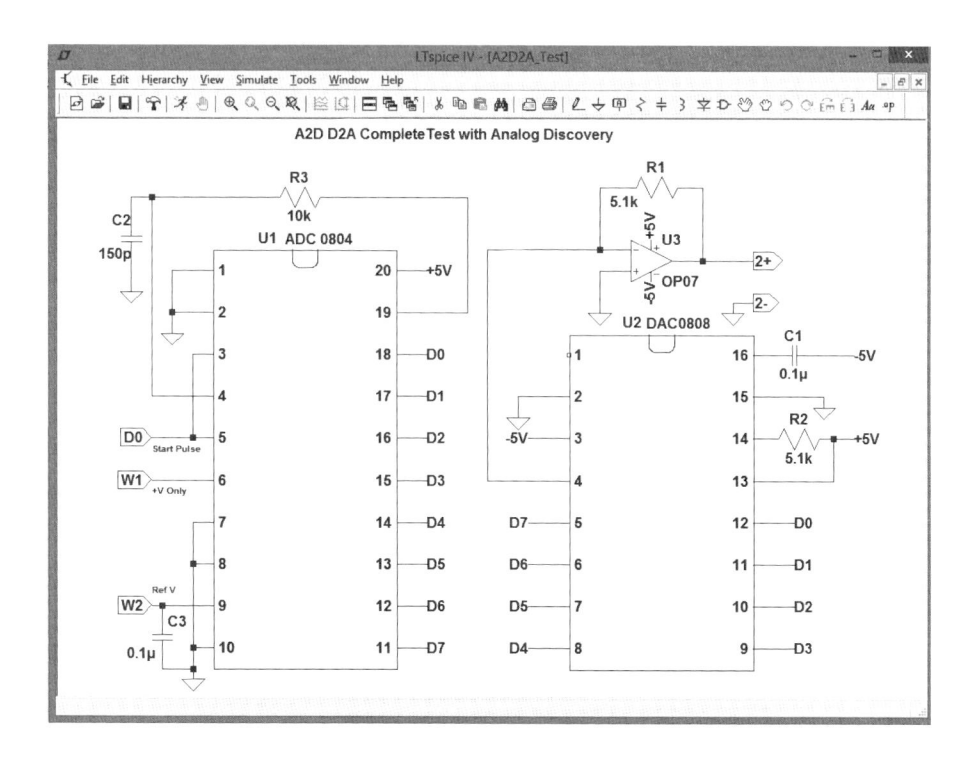

圖 8-4　ADC0804 - DAC0808 的聯合測試電路

電路和 Analog Discovery 之間的連線如下列所示：

W1	→	U4 pin3
W2	→	U2 pin9
+5 V	→	U1 pin14，U2 pin20，U3 pin7，U4 pin7。
−5 V	→	U1 pin3，U3 pin4，U4 pin4。
GND	→	GND
2+	→	U3 pin6
2−	→	GND
DIO 0	→	U2 pin3，U2 pin5。

聯合測試電路，事實上是第七章，圖 7-10 的 DAC 電路 + ADC 電路。圖 8-5 是它在聯合測試電路在麵包板上的佈置。圖 8-6 為 Analog Discovery 與電路的連線。

Analog Discovery 的 W1 提供 2 V、1 Hz 正弦波的測試輸入。W2 提供 ADC0804 pin 9 的直流 2.5 V 的 REF 電壓。Pattern Generator DIO 0 提供 pin 3 和 pin 5 所需的觸發和輸出命令訊號。

圖 8-7 是在 WaveForm 中選用電源。圖 8-8 WaveForm 中選用 **Analog** 的 Function Generator W1 和 W2，這些都是前面幾章中所常用的。

◌ 圖 8-5 ◌　　　聯合測試電路在麵包板上的佈置

第八章　類比訊號轉換成數位訊號　87

圖 8-6　Analog Discovery 與電路的連線

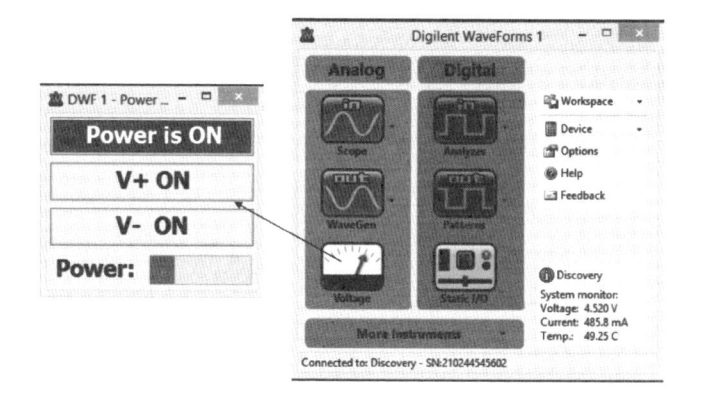

圖 8-7　WaveForm 中選用電源

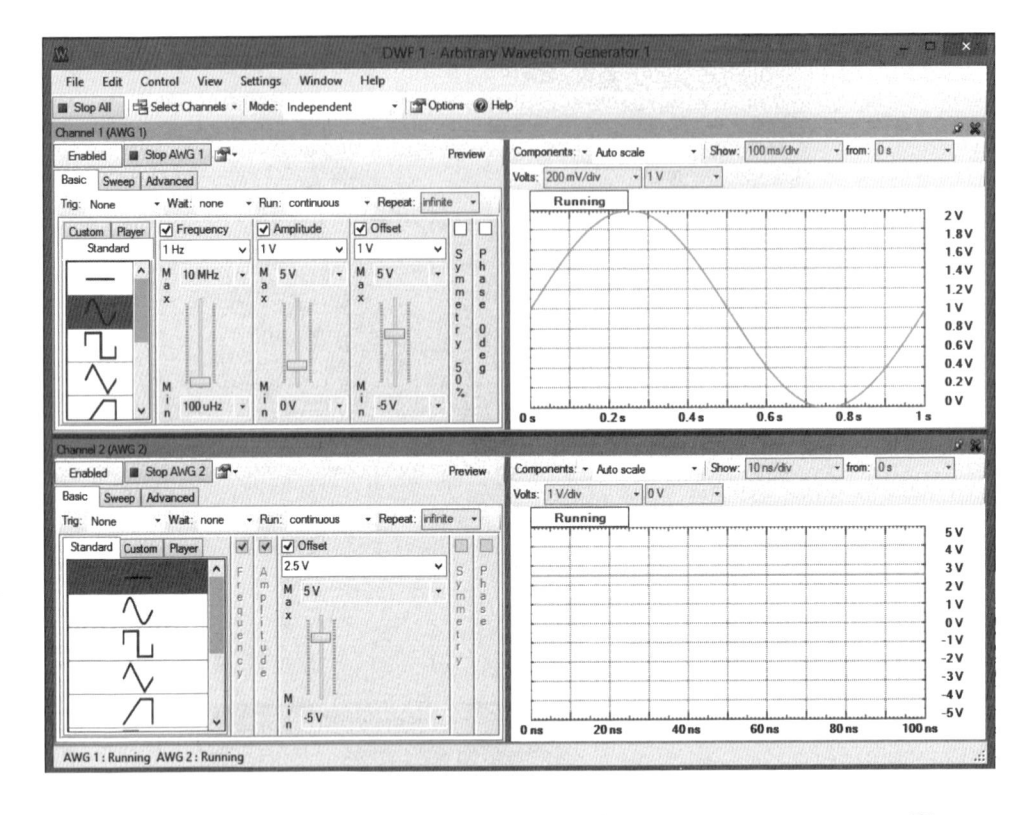

圖 8-8　　WaveForm 中選用 Analog 的 Function Generator W1 和 W2

　　Pattern Generator DIO 0 提供 pin 3 和 pin 5 所需的觸發和輸出命令訊號。要從圖 8-7 WaveForm 中選用 **Digital** 的 **OUT** 入手，如圖 8-9 所示。

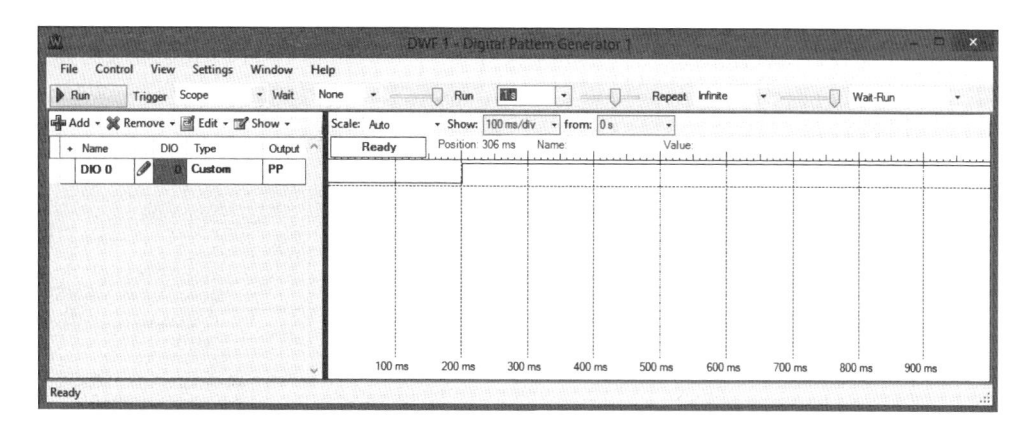

第八章　類比訊號轉換成數位訊號　89

◦ 圖 8-9 ◦　　　WaveForm 中選用 Digital 的 Pattern Generator

　　圖 8-10 為 Pattern Generator 的設定，**Type** 選用 **Customer**，**Output** 選用 **PP**，**Idle** 選用 **High**，**Buffer** 選用 **1024**，Frequency 為 1 KHz。在 Prefill 項的 **Type** 應選用 **Pulse**，**Low** 為 200，**High** 為 824。

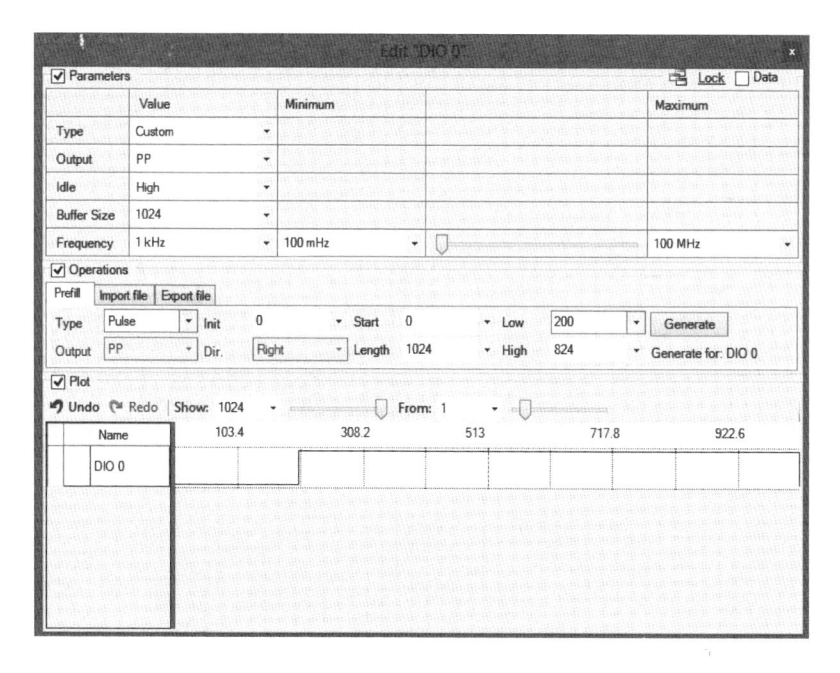

◦ 圖 8-10 ◦　　　Pattern Generator 的設定

如果電路的接線沒錯，設定也是對的，則單擊圖 8-11 示波器 2+ 的 RUN，應當觀察到一個相似於圖 8-8 W1 所輸入的正弦波。

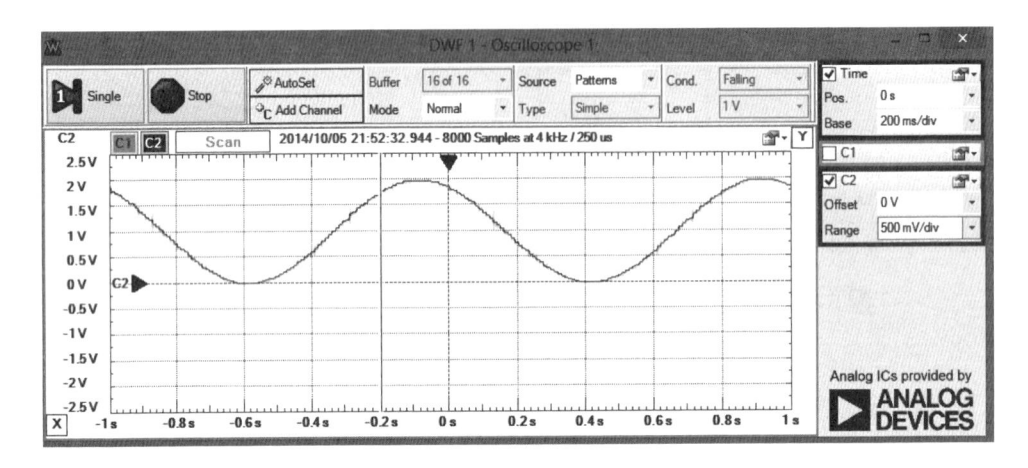

◎ 圖 8-11 ◎　　　示波器 2+ 測試 ADC－DAC 輸入訊號的還原

8-4　課外練習

1. 試根據圖 8-3 說明簡易測試 ADC0804 所需的步驟。

2. 試說明 ADC 的精確度與其組成 bits 多寡的關係。

3. 圖 8-4 的 ADC0804-DAC0808 聯合測試電路，所加入之電壓為圖 8-8 AWG1 的 0 V 至 +2 V 之正弦波。若改變為 –1 V 至 +1 V 之正弦波，如果要得到正確的還原輸出，試說明圖 8-4 的測試電路應做哪些改變？並用實作證明之。

4. 試從 Analog Discovery Module 的資料中找出它用的是何種 ADC？由多少個 bits 所組成？Sampling rate 是多少？能測試到正弦波的最高頻率是多少？

第九章　Clock Generation 與 PLL

時序脈波是推動數位系統不可或缺的元件。**鎖相環** (Phase Locked Loop) 是在一連串輸入訊號中掏取 Clock 的電路。Clock 的產生和 PLL 是本章要涉及的主題。

9-1　簡單時序脈波的產生

圖 9-1 為簡單**時序脈波** (Clock) 的產生器電路，使用數位系統中二個剩餘的 inv 如 74LS04 或 nand，nor gates 串聯回授，中間再加上控制 Clock 頻率的石英晶體即可完成。Clock Out 所產生的頻率，也就是石英晶體的自然頻率，方波的強度依 Gate 的種類而定。

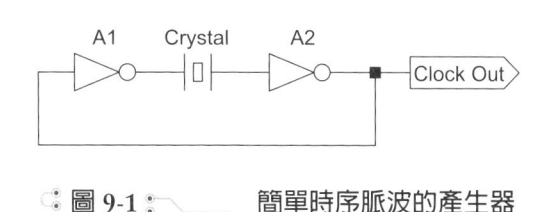

圖 9-1　簡單時序脈波的產生器

9-2　Clock 的分佈

Gate 的輸出有一定的限制，解決的方法是用多個 Gates 來展開，如圖 9-2 所示。要注意的是，輸出的相位和 Gates 在時間上的延遲。

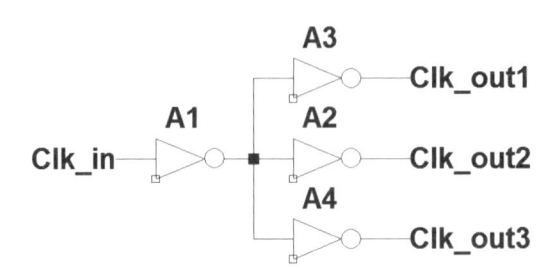

圖 9-2　用 Gates 展開來解決輸出限制的電路

9-3　不同頻率的要求

系統中可能有不同頻率的要求，如果是低於原始 Clock 的頻率，可以用 Counter 的除法來降低而獲得。如果是高於原始 Clock 的頻率，則需用較複雜的 PLL 電路來獲得。

9-4　鎖相環 PLL 電路的結構

頻率鎖相電路是由四個部份所組成，如圖 9-3 所示，其中的相位檢測器大多屬於數位電路外，其它的低通濾波器、放大器、和電壓控制振盪器，都是屬於類比電路。頻率鎖相電路，主要是用在通信電子上，從一連串的數位的 '0' 和 '1' 不規則的信號中，撈取 Clock 信號。

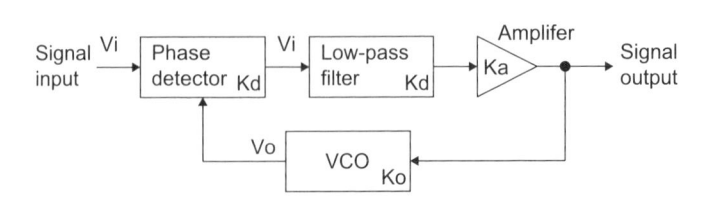

圖 9-3　頻率鎖相電路 PLL 的電路組成

9-5 相位檢測器

數位訊號的相位檢波電路種類甚多，最為常用的是圖 9-4 被用於 PLL 的相位和頻率檢波電路，它的 VHDL 模式如圖 9-5 所示。ModelSim 的 TestBench 如圖 9-6。有關 VHDL 電路和 Testbench 的構成，請參考附錄 A 和 B。ModelSim 的 Simulation 步驟請參考附錄 E。測試結果如圖 9-7 和 圖 9-8。

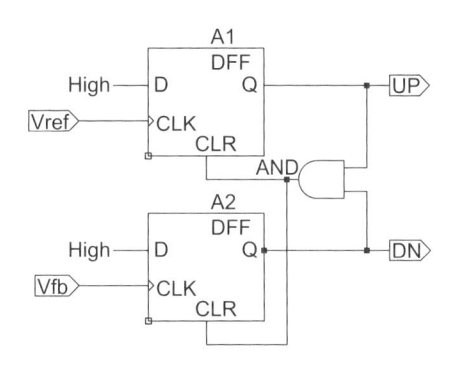

圖 9-4　用於 PLL 的相位和頻率檢波電路

```
1  ENTITY PhaseDetector IS
2  PORT ( Vref, Vfb: IN BIT;
3                   clr: INOUT BIT;
4                   UP, DN: INOUT BIT);
5  END ENTITY;
6
7  ARCHITECTURE arch OF PhaseDetector IS
8     Component DFF
9     PORT ( d, clk, clr: IN BIT;
10                  q: OUT BIT);
11    END Component;
12
13    Component AND2
14      PORT ( a, b: IN BIT;
15                  c: OUT BIT);
16    END Component;
17
18  Begin
19
20    A1:  DFF port map ( '1', Vref , clr, UP);
21    A2:  DFF port map ( '1', Vfb,  clr, DN);
22    A3:  AND2 port map ( UP, DN,  clr);
23
24  END ARCHITECTURE;
```

圖 9-5　相位和頻率檢波器的 VHDL 模式

```
1  Library ieee;
2  Entity Tstbench IS
3  END Tstbench;
4  USE work.all;
5
6  Architecture stimulus of Tstbench IS
7
8     Component PhaseDetector
9     PORT ( Vref, Vfb: IN BIT;
10                  clr: INOUT BIT;
11              UP, DN: INOUT BIT);
12    END Component;
13
14  SIGNAL Vref, Vfb, clr: BIT := '0';
15  SIGNAL UP, DN: BIT := '0';
16  Begin
17    DUT: PhaseDetector port map ( Vref, Vfb, clr, UP, DN);
18
19    Vref <= NOT Vref After 110 ns;
20    Vfb  <= NOT Vfb  After 100 ns;
21  END stimulus;
```

圖 9-6　相位和頻率檢波器的 VHDL 測試模式

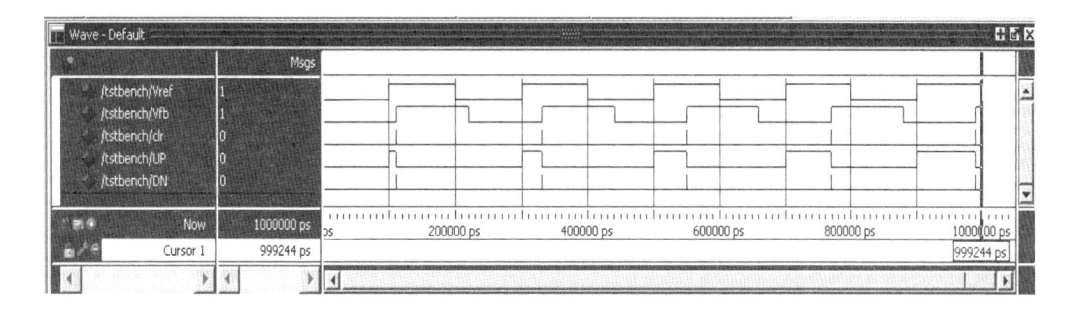

◎ 圖 9-7 ◎ ～～ Vref 頻率高於 Vfb 時，檢波器的 UP 有正向脈波輸出

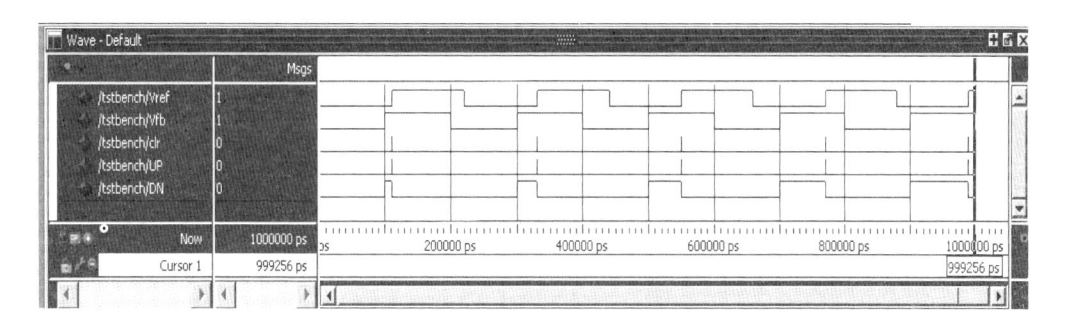

◎ 圖 9-8 ◎ ～～ Vref 頻率低於 Vfb 時，檢波器的 DN 有正向脈波輸出

　　從圖 9-7 可以看到當 Vref 頻率高於 Vfb 時，檢波器的 UP 有正向脈波輸出。而當 Vref 頻率低於 Vfb 時，如圖 9-8，則檢波器的 DN 有正向脈波輸出。這個 UP 的正向脈波將經過 Low Pass Filter 和放大器之後輸入到 VCO，讓 Vfb 的輸出頻率升高。而 DN 的正向脈波將經過 Low Pass Filter 和放大器，並轉變為負壓之後，輸入到 VCO 讓 Vfb 的輸出頻率降低。

　　Low Pass Filter，Amplifier 和 VCO 全部是 Analog 的範圍。LM565 的設定請參考附錄 F。

9-6　PLL 的 Clock 頻率倍增器

圖 9-3 頻率鎖相電路 PLL 的電路中，如果在 VCO 和 Phase detector 間加入一個除以 N 的計數器，如圖 9-9 所示，則 VCO 的輸出 V0，其頻率當為 Signal input Vi 的 N 倍。

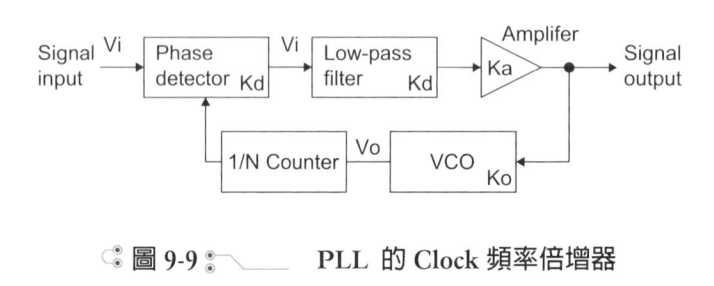

圖 9-9　PLL 的 Clock 頻率倍增器

9-7　LM565/PLL 和 74LS90/Decade Counter 構成的 ×10 倍頻器

圖 9-10 是 LM565 構成的 10× 倍頻電路，10× 的獲得是讓 VCO 的輸出，不直接連到相位檢波器。而是連接到 74LS90，這個 ÷10 的**計數器**(Counter) 上，再把 VCO ÷10 後的 f_0，拿去跟外來的 10 KHz 訊號做相位檢波，鎖相的結果，VCO 當然必須為 10 × 10 KHz = 100 KHz 了。

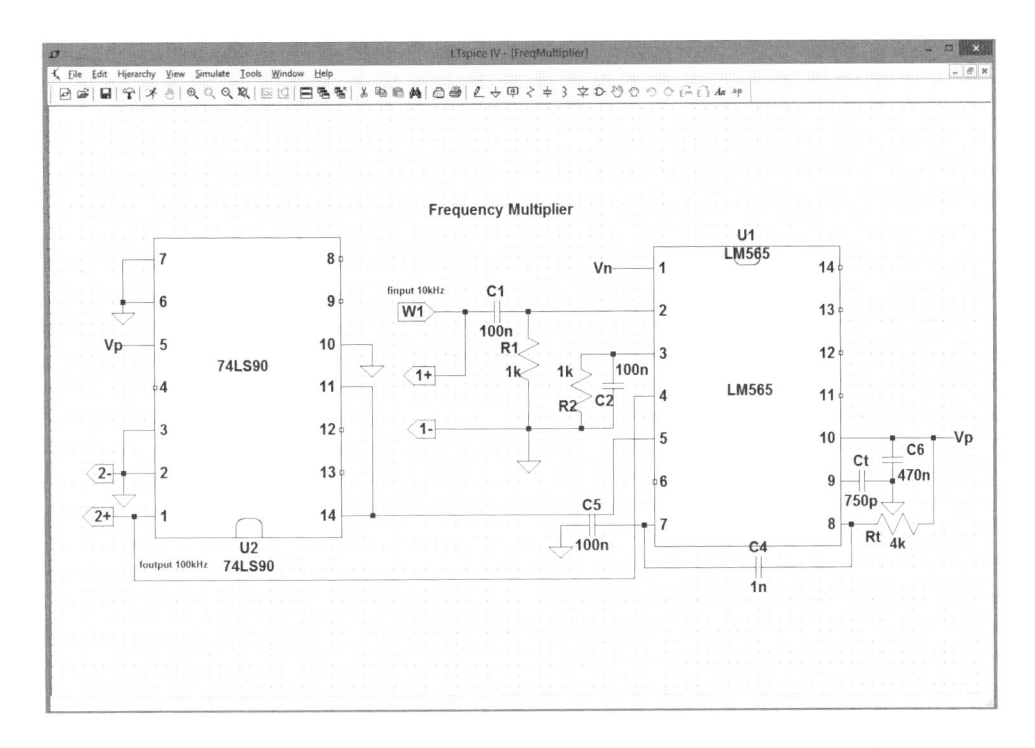

第九章 Clock Generation 與 PLL　99

圖 9-10　LM565 和 74LS90 構成的 ×10 倍頻器電路

9-8 硬體實作

由於 LM565 並無 Spice Model 的提供，圖 9-10 就是本電路與 Analog Discovery Module 的接線關係圖。簡單的低頻電路，如果只為了學習，可以使用不用銲接的**麵包板** (Breadboard) 來完成。頻率較高或比較複雜的電路，應當把電路裝配在接地良好，使用銲接的電路板上。如圖 9-11 所示。

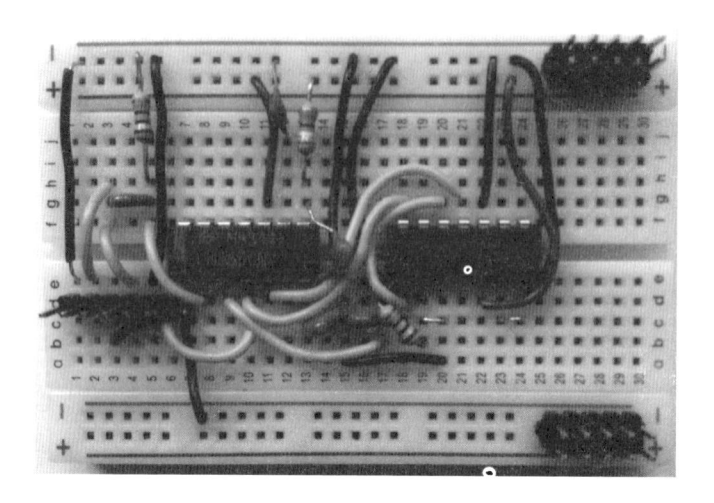

圖 9-11　PLL 的頻率乘 10 電路的麵包板實作裝置圖

　　Analog Discovery Module，連同裝上了 WaveForm 軟體的 PC，形成了多件由軟體操控的硬體儀器。點擊 Analog 部份的 **Voltage**，如圖 9-12 所示。這 2 個 5 V 電源，可以 ON 或 OFF 軟體控制。

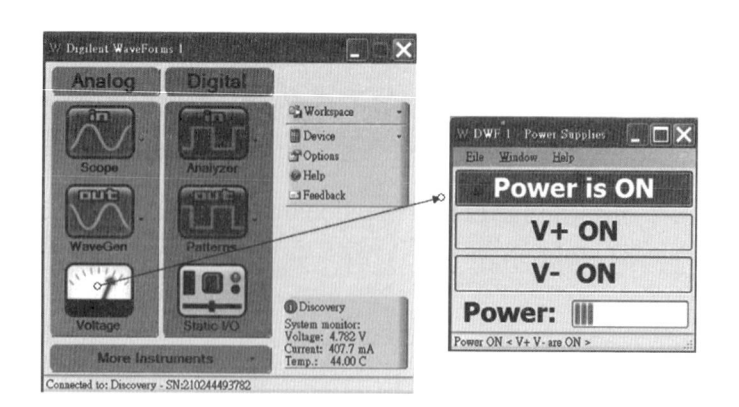

圖 9-12　Analog Discovery 的 2 個可 ON 或 OFF 的電源

PLL 的輸入，由 W1 提供，它是一個電壓為 3.5 V 的方波，如圖 9-13 所示。

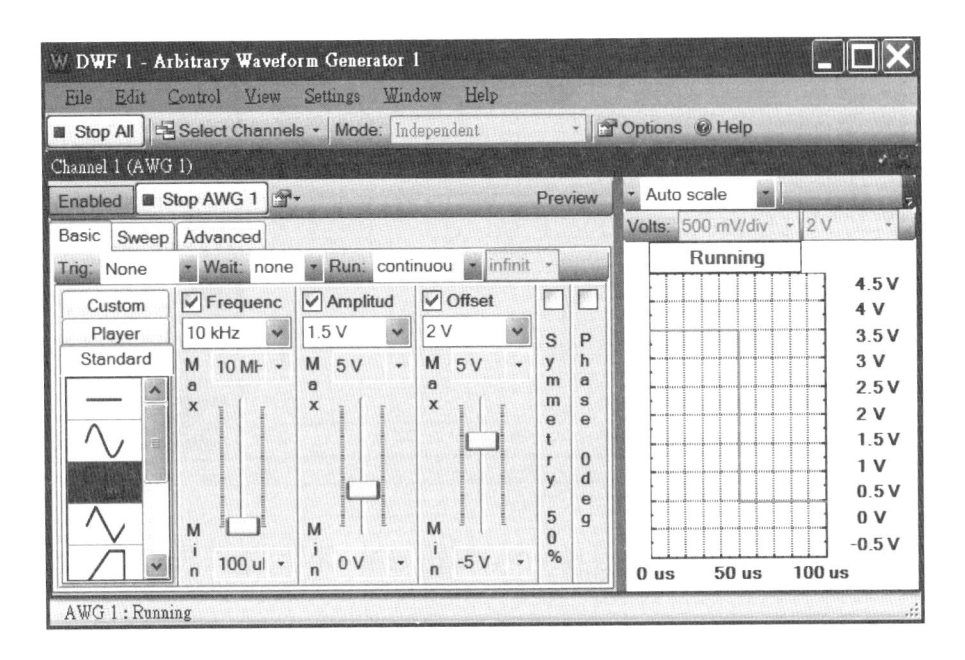

図 9-13 輸入到 PLL 的訊號為 3.5 V 的方波由 W1 提供

示波器 1+ 連接到 PLL 的輸入端，2+ 連接到 10× Counter 的 PLL 端。1- 和 2- 接地。當鎖相發生時，Counter 的輸入頻率，為 W1 輸入頻率的 10 倍。如圖 9-14 所示。

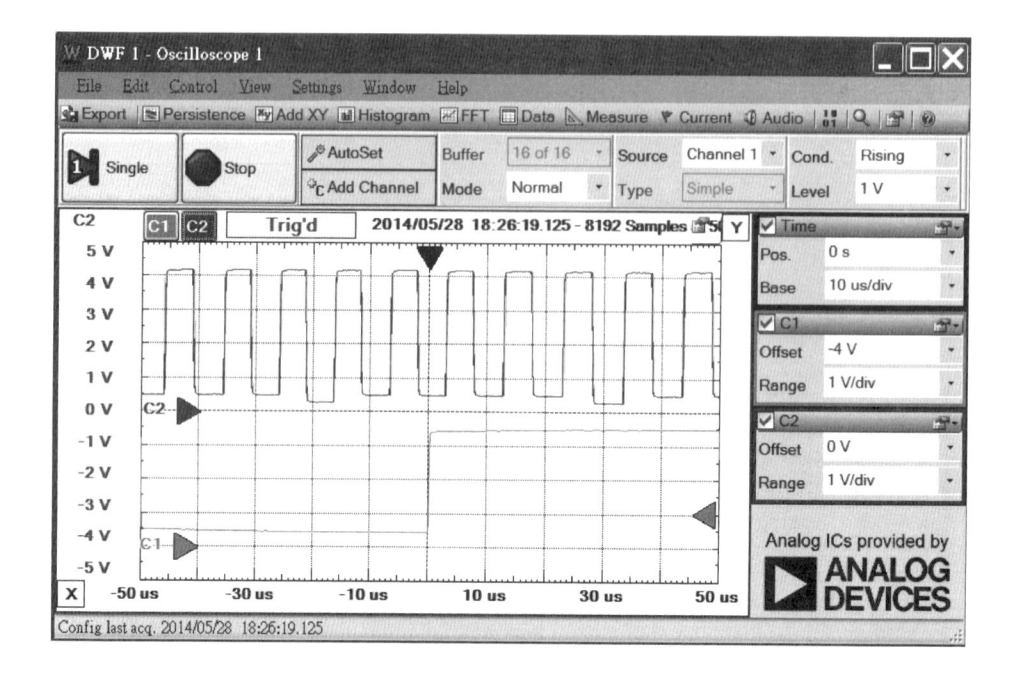

圖 9-14　鎖相發生時 VCO 的輸出頻率，為 W1 輸入頻率的 10 倍

9-9 課外練習

1. 試用 SN74LS74 和 SN74LS08 來組成圖 9-4 用於 PLL 的相位和頻率檢波電路，並用 Analog Discovery 來測試之。

2. 試用 SN74LS04 和 4 MHz 的石英振盪器來組成圖 9-1 和圖 9-2 的合併電路，並用 Analog Discovery 來測試之。

3. 試用題 2. 所產生之 4 MHz 來產生 1 MHz 和 100 KHz Clock 頻率。

4. 試用題 2. 所產生之 4 MHz 來產生二個 1 MHz/0 度和 1 MHz/90 度頻率的訊號。

第十章　Step Motor 與 Driver

　　Step Motor 是數位電路專屬的馬達，以線圈組成的不同，可以分成 Unipolar 和 Bipolar 二大類。它和一般直流和交流馬達最大的不同點，是以脈波來推動。能精確地做或快或慢的順轉和逆轉運動。是系統設計中對於位置控制的重要元件。

10-1　單相 Unipolar 步進馬達

　　單相步進馬達實際上有 A、B、C、D 共 4 組線圈。推動的方式：分別為單相、雙相和半步等三種。

(A) **單相** (Single-phase) 推動方式為每次以勵磁一個線圈為原則，序列的時間為 ABCD (或 DCBA)，如圖 10-1 所示：

Step	A	B	C	D
POR	ON	OFF	OFF	OFF
1	ON	OFF	OFF	OFF
2	OFF	ON	OFF	OFF
3	OFF	OFF	ON	OFF
4	OFF	OFF	OFF	ON

圖 10-1　步進馬達 4 組線圈單向勵磁序列

這種勵磁方式消耗最少的功率,並保證定位的精確度。

(B) **雙相** (Two-phase) 推動方式為每次以勵磁二個相鄰線圈為原則,序列的時間為 (AB-BC-CD-DA)。如圖 10-2 所示:

Step	A	B	C	D
POR	ON	OFF	OFF	ON
1	ON	OFF	OFF	ON
2	ON	ON	OFF	OFF
3	OFF	ON	ON	OFF
4	OFF	OFF	ON	ON

圖 10-2　　步進馬達 4 組線圈雙相勵磁序列

這個序列模式提供更高的起動轉矩,並且不容易引起馬達發生共振。

(C) **半步** (Half-step) 推動方式為單階段和兩階段模式 (A-AB-B-BC-C-CD-D-DA) 之間的中間步驟交替激發。如圖 10-3 所示:

Step	A	B	C	D
POR	ON	OFF	OFF	OFF
1	ON	OFF	OFF	OFF
2	ON	ON	OFF	OFF
3	OFF	ON	OFF	OFF
4	OFF	ON	ON	OFF
5	OFF	OFF	ON	OFF
6	OFF	OFF	ON	ON
7	OFF	OFF	OFF	ON
8	ON	OFF	OFF	ON

圖 10-3　　步進馬達 4 組線圈半步勵磁序列

半步推動方式,提供一個 8 個步驟的序列。

10-2 Bipolar 步進馬達

Bipolar 步進馬達只有 2 組線圈,為了達到 4 組線圈的同樣效果,須用橋狀供電,如圖 10-4 所示。

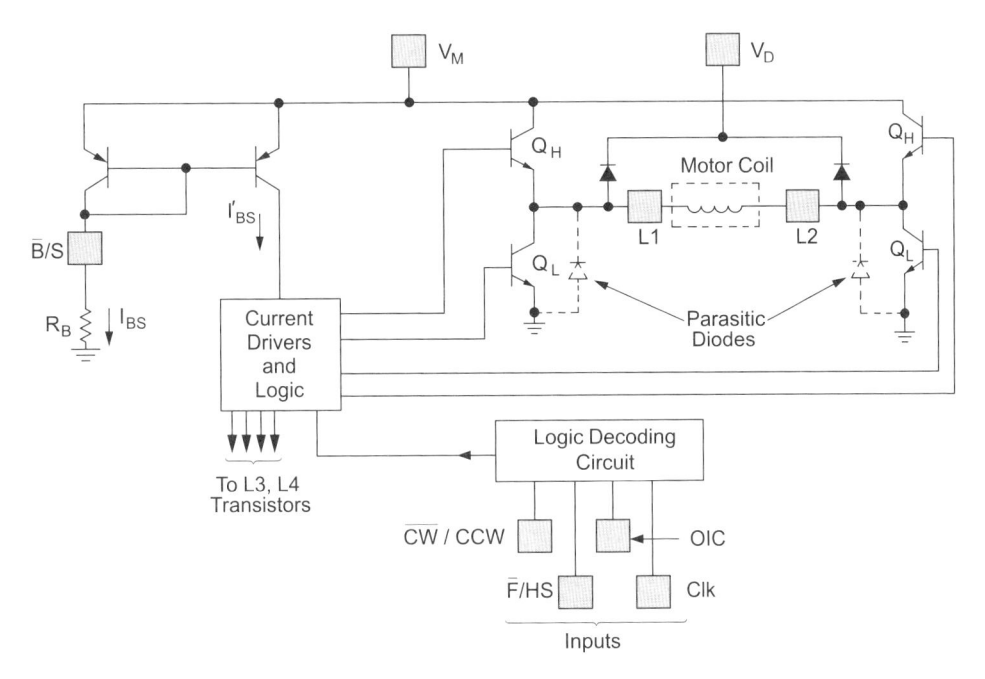

圖 10-4 MC3479 IC 橋狀供電設計以推動 2 組線圈的步進馬達

10-3 5804 IC 以推動 4 組線圈的步進馬達

圖 10-5 為 5804 IC 的推動 4 組線圈的步進馬達電路,它不但能提供三種不同模式的選擇,並且還提供步進馬達電功率所需之 BiMOS 電晶體。

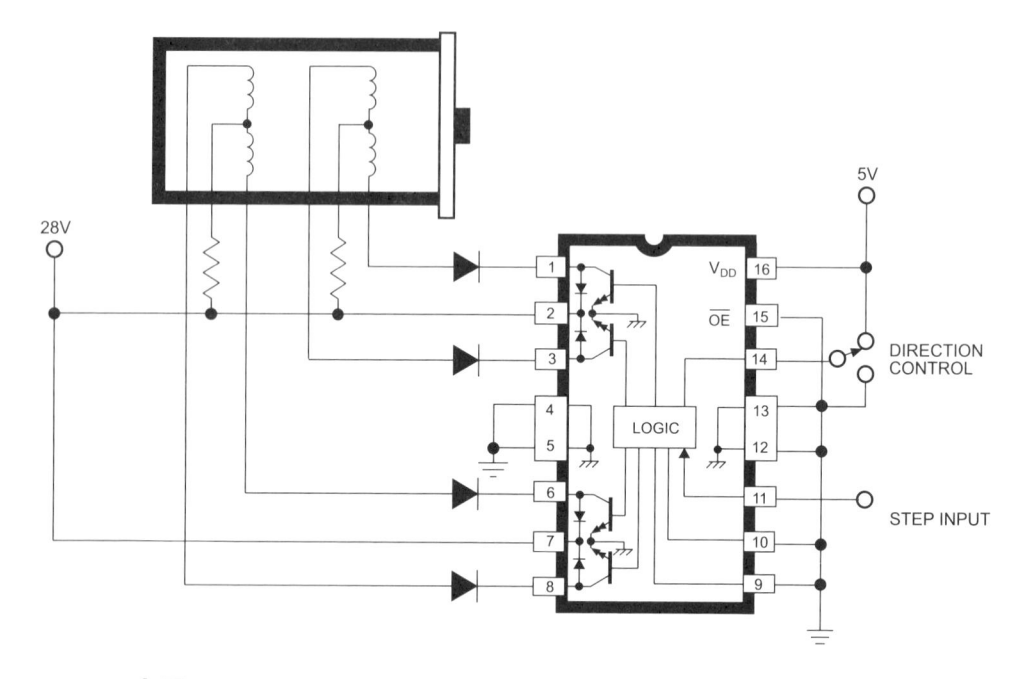

圖 10-5　5804 Unipolar Step motor translator and driver

10-4　微小型 Unipolar 步進馬達的規格

下面所列出的是 28BYJ-48，用在 ARDUINO 上的微小型 Unipolar 步進馬達的規格：

Electronic Parameters：
- Rated voltage：5VDC
- Number of Phase：4
- Speed Variation Ratio：1/64
- Stride Angle：5.625°/64
- Frequency：100Hz
- DC resistance：50Ω±7%(25℃)
- Idle In-traction Frequency：> 600Hz
- Idle Out-traction Frequency：> 1000Hz
- In-traction Torque >34.3mN.m(120Hz)
- Self-positioning Torque >34.3mN.m
- Friction torque：600-1200 gf.cm
- Pull in torque：300 gf.cm
- Insulated resistance >10MΩ(500V)
- Insulated electricity power：600VAC/1mA/1s
- Insulation grade：A
- Rise in Temperature <40K(120Hz)
- Noise <35dB(120Hz,No load,10cm)
- Model：28BYJ-48

10-5 硬體實作

圖 10-6 為 Analog Discovery 對 28BYJ-48 Unipolar 步進馬達的測試連線，馬達的驅動板用的是 ULN2003 [1]。

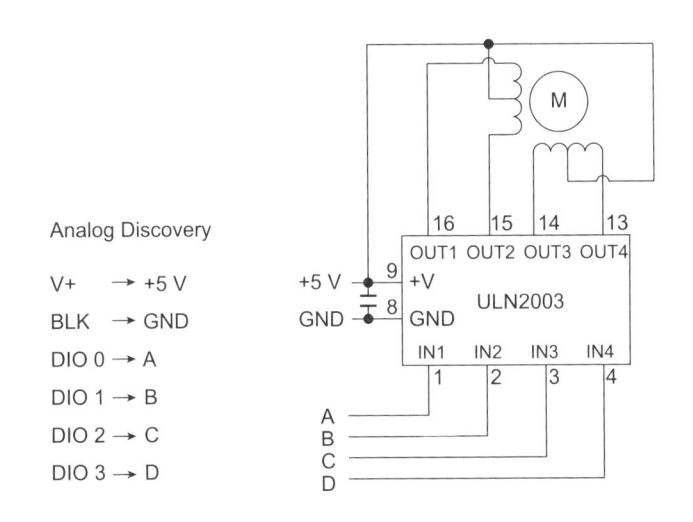

圖 10-6　Analog Discovery 對步進馬達的測試連線

---

[1] UL2003 馬達的驅動板 +5 V 步進馬達為 ARDUINO 配件 MTARDUL2003MO (台北光華商場)。

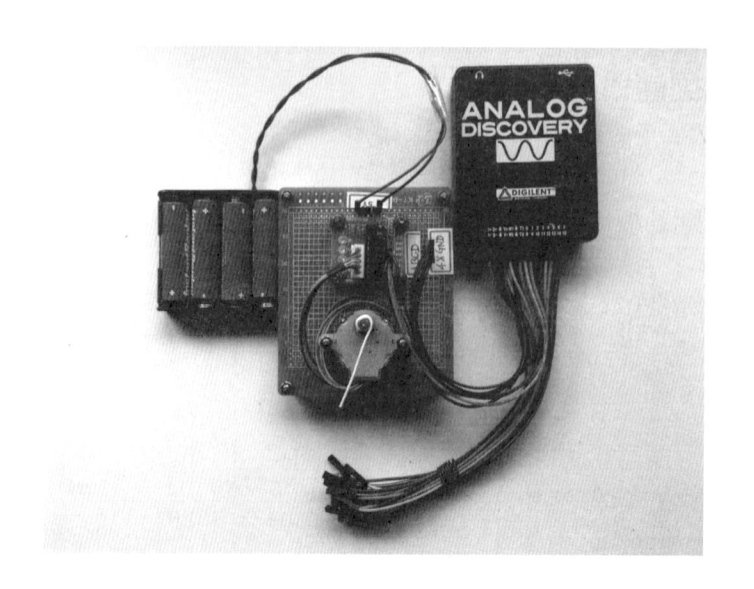

圖 10-7　ULN2003 驅動板與步進馬達及 Analog Discovery 直接連接

　　如果使用的是 ULN2003 驅動板，上面有插針可以與 Analog Discovery 直接交連，還有插座供步進馬達之用，所以完全不需用麵包板。

　　使用 Analog Discovery Module，連同裝上了 WaveForm 軟體的 PC，成為多件由軟體操控的硬體儀器。測試數位電路，對於電路的輸入部份可用**數字模式產生器** (Digital Pattern Generator)，輸出部份可用**數位信號分析儀** (Digital Signal Analyzer)，圖 10-8 就是 WaveForms 只在 PC 上所顯示的選擇圖形。

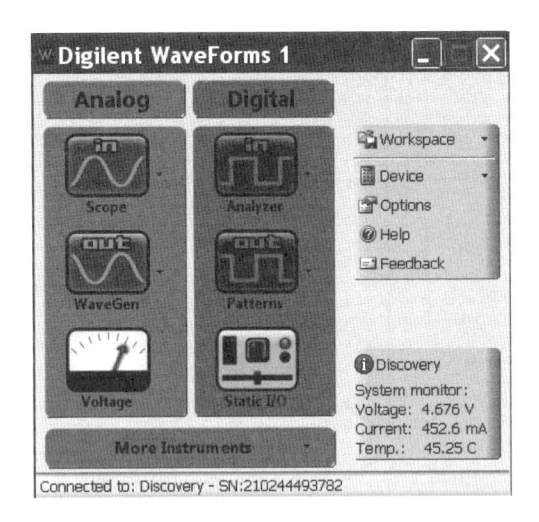

圖 10-8　　WaveForms 只在 PC 上所顯示的選擇圖形

　　步進馬達的測試，也跟測試一般邏輯電路相仿。選用 **Digital** 部份的 OUT 用 Logic Pattern Generator 來產生輸入訊號。輸出訊號對於步進馬達來講是馬達轉動軸的旋轉，Analog Discovery 並沒有直接測試轉動軸旋轉的設施，所以只有憑個人的目視觀察了。

　　由於 ULN2003 的 INV Logic 如圖 10-9 所示，因此在設定勵磁序列時要將圖 10-1 步進馬達 4 組線圈單向勵磁序列圖表上的 ON 設成 OFF，OFF 設成 ON。如果 Analog Discovery 的 DIO_0～DIO_3 連接到 ULN2003 的 IN1～IN4，也就是 A、B、C、D。其邏輯訊號，當如下列所示。

DIO_0 → A : 1, 0, 1, 1, 1
DIO_1 → B : 1, 1, 0, 1, 1
DIO_2 → C : 1, 1, 1, 0, 1
DIO_3 → D : 1, 1, 1, 1, 0

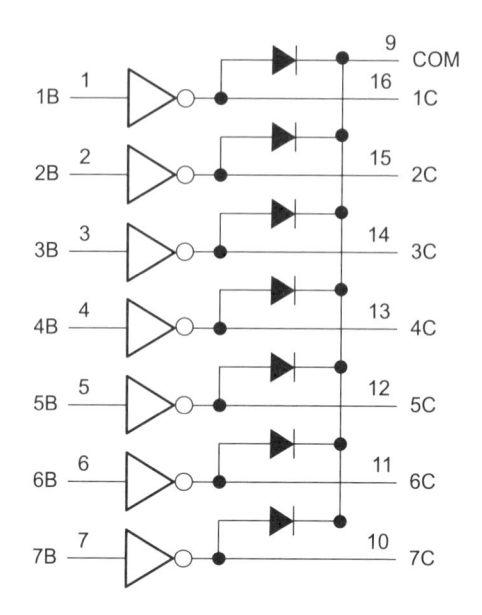

◎ 圖 10-9　　　Logic Diagram of ULN2003

　　由圖 10-6 ULN2003 的輸入訊號共 4 個 A、B、C、D，Digital Pattern Generator 提供 DI0 0～DIO 3 相對應，如圖 10-10 所示。

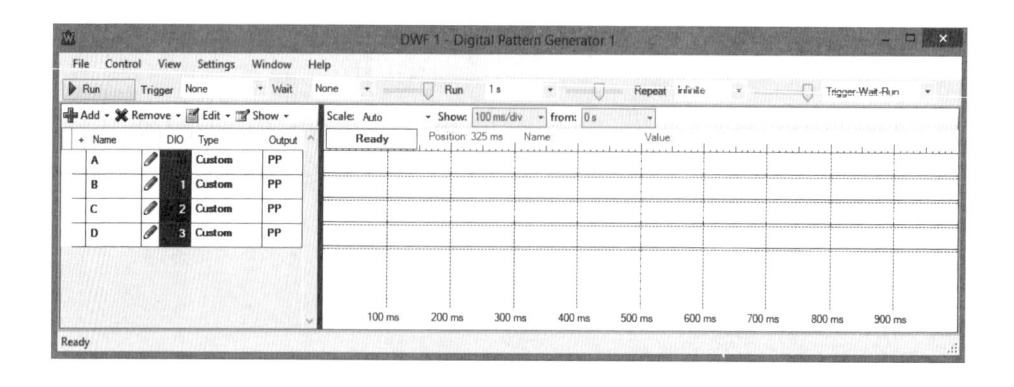

◎ 圖 10-10　　　DIO_0～DIO_3 連接到 ULN2003 的 Pin 1、2、3、4

第十章　Step Motor 與 Driver　113

　　DIO_0~DIO_3 的設定，除了勵磁序列外，其他完全相同。圖 10-11 是 DIO_0 的一個例子，Parameters 中的 **Type** 為 **Customer**，**Output** 為 **PP**，**Idle** 是依圖 10-1 的 **POR** 而定，如圖 10-11 DIO 0 的 **Idle** 為 **Low**，圖 10-12 DIO_1~DIO_3 則為 **High**。Buffer Size 為 4，Frequency 可訂為 10 Hz。波形設定完了之後，須按右上角的 LOCK 加以固定。

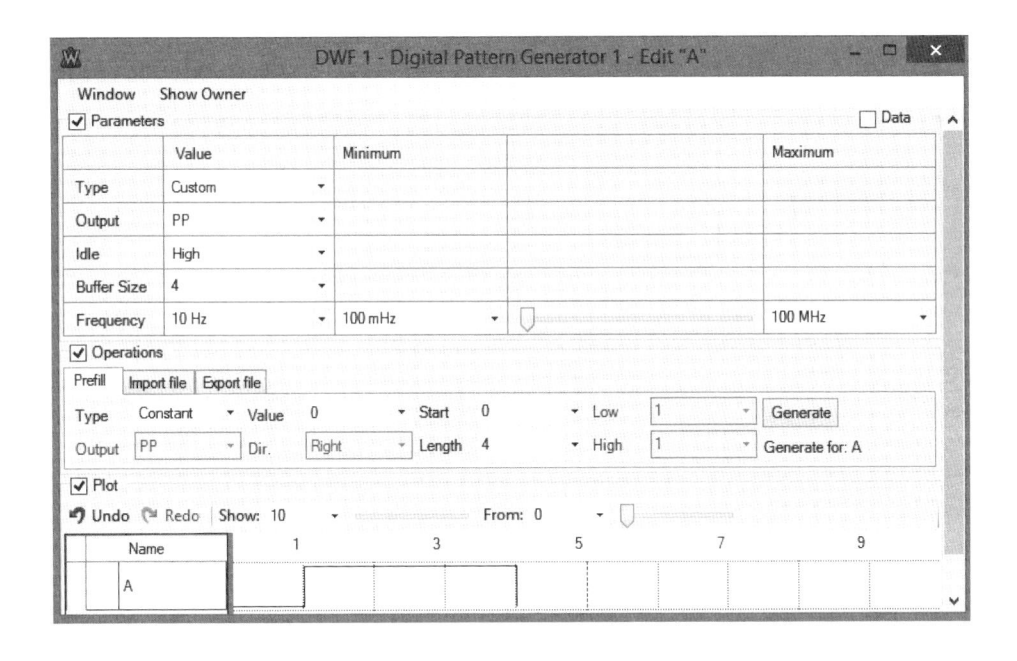

◖ 圖 10-11 ◗　　DIO 0 連接到 ULN2003 IN1 的設定

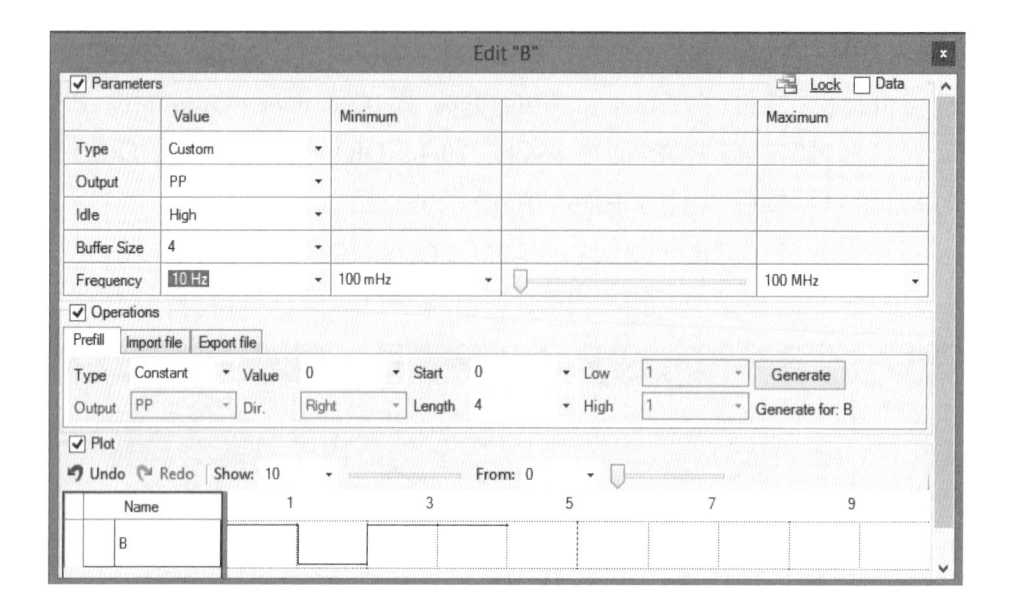

◉ 圖 10-12 ◉　　DIO 0 連接到 ULN2003 IN2～IN4 的設定

全部設定完了再 RUN 時，Digital Pattern Generator 的輸出波形當如圖 10-13 所示。

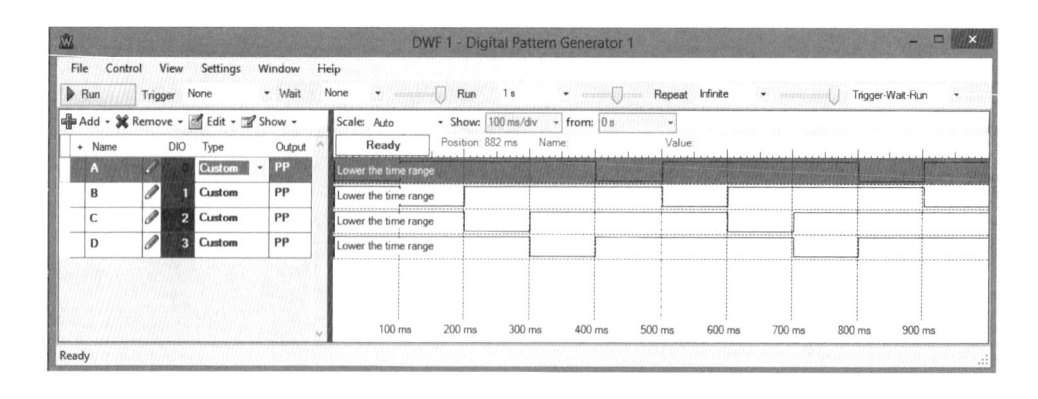

◉ 圖 10-13 ◉　　Digital Pattern Generator 的輸出到 ULN2003 的波形

Analog Discovery 的 ±5 V 電源，最多只能提供 50 mA 的電流。步進馬達 28BYJ-48 從第 10-4 節的規格，其線圈的直流電阻爲 50 Ω。直流瞬間電流 I = 5/50 = 100 mA，因此不能使用該電源來供電。圖 10-7 使用的 +5 V 外接電源，爲了 Module 的安全，是由 4 只 1.25 V NI-MH 電池串連而成。注意電池的負端，必須跟 Module 的 4 條 GND (黑線) 連接起來。

10-6　課外練習

1. 硬體實作中使用的是圖 10-2 的步進馬達 4 組線圈雙相勵磁序列。試改變為圖 10-3 的步進馬達 4 組線圈半步勵磁序列，比較二者的異同點。

2. 硬體實作中使用的是圖 10-2 的步進馬達 4 組線圈雙相勵磁序列，結果是順時鐘旋轉。試寫出逆時鐘旋轉之雙相勵磁序列。

3. 在沒有 CPU 的情況下，如何能達成順時鐘旋轉和逆時鐘旋轉的功能？

第十一章　Servo System 與 Control

伺服系統與步進馬達不同,它是一個回授系統:其中包含著一個小型電動馬達,通過齒輪連接到輸出軸上。由輸出軸驅動的伺服臂,並且還連接到一個可變電阻的電位器,以提供伺服臂位置的訊息,反饋到內部控制電路。伺服系統可以準確地控制,其旋轉範圍一般為 0°～180°。常被用於受遙控的機械如生產線上的傳送帶、飛機機翼和直升機升降等之控制。

11-1　伺服系統的內部結構

伺服系統的內部結構如圖 11-1 所示。

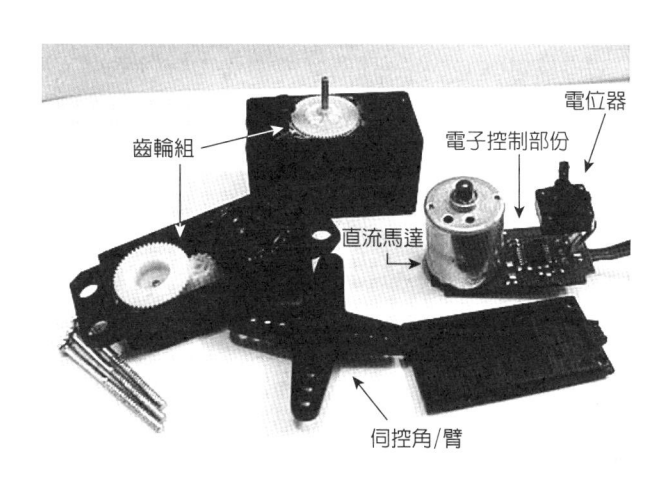

圖 11-1　伺服系統的內部結構

它的電子系統方塊圖如圖 11-2 所示。

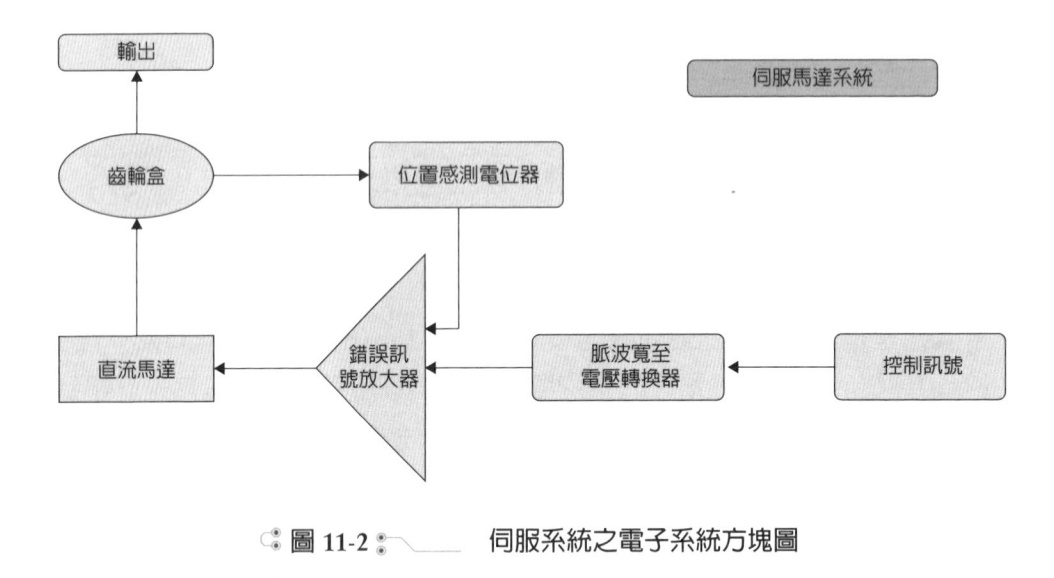

▓ 圖 11-2 ▓　　伺服系統之電子系統方塊圖

11-2　伺服系統的控制

伺服馬達可以由控制輸入端加以**脈衝寬度調變** (Pulse Width Modulation, PWM) 信號，使移動到所希望的角度位置。脈衝以 50 Hz 的頻率，寬度從 1 到 2 ms 的持續變化，來決定並保持角度的位置，如圖 11-3 所示。

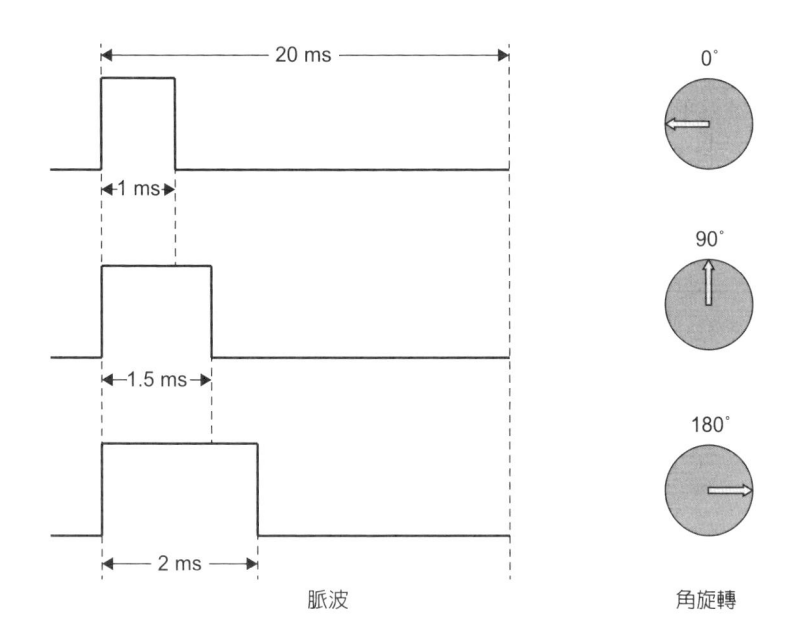

圖 11-3　伺服系統角度位置與輸入脈衝寬度調變信號的關係

11-3　Micro Servo MG90S [1] 的規格與連線

圖 11-4 是一個超小型的伺服系統 MG90S 的規格。圖 11-5 為其連線。

---

[1] MG90S 伺服馬達為 ARDUINO 配件料號 MOMG90S (台北光華商場)。

MG90S
Metal Gear Servo

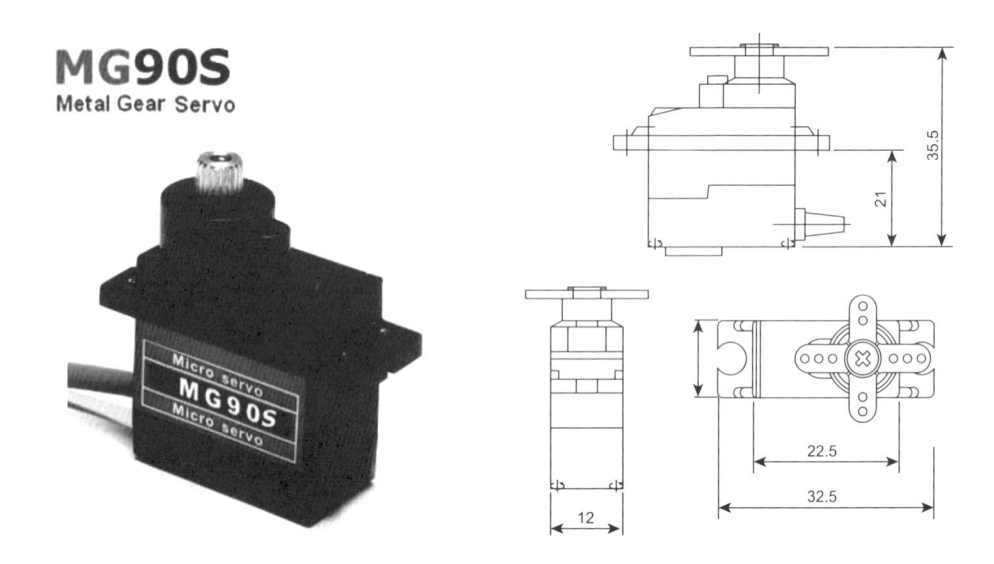

Specifications

- Weight: 13.4 g
- Dimension: 22.5 x 12 x 35.5 mm approx.
- Stall torque: 1.8 kgf.cm (4.8 V), 2.2 kgf.cm (6 V)
- Operating speed: 0.1 s/60 degree (4.8 V), 0.08 s/60 degree (6 V)
- Operating voltage: 4.8 V – 6.0 V
- Dead band width: 5 µs

圖 11-4　　超小型伺服系統 MG90S 的規格

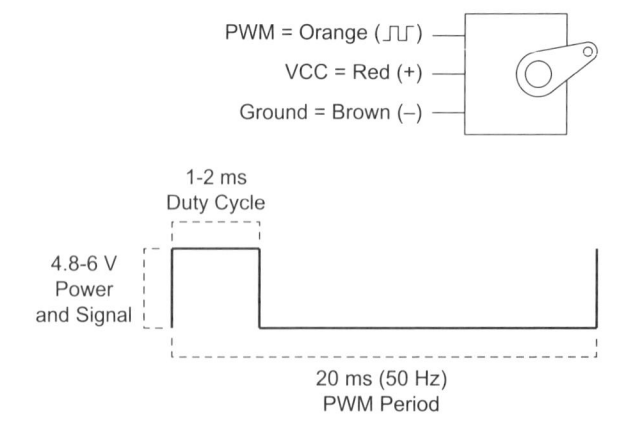

PWM = Orange (⎍)
VCC = Red (+)
Ground = Brown (−)

1-2 ms Duty Cycle
4.8-6 V Power and Signal
20 ms (50 Hz) PWM Period

圖 11-5　　超小型伺服系統 MG90S 的連線

11-4　Servo Control IC

　　積體電路的伺服系統有 NE544 和比較新型的 M51660 [2] 等多種，它們接受同樣的脈波輸入，利用控制信號來控制旋轉的角度，但消除了系統中使用機械式的可變電阻。使用外接電晶體，可接用較大電流的 Servo，同時可對觸發訊號及 Deadband 加以調整。電路的內部還有穩壓的設置，所以比起舊式的 MG90S 較為優良。圖 11-6 便是使用 M51660 的 Servo 電路。

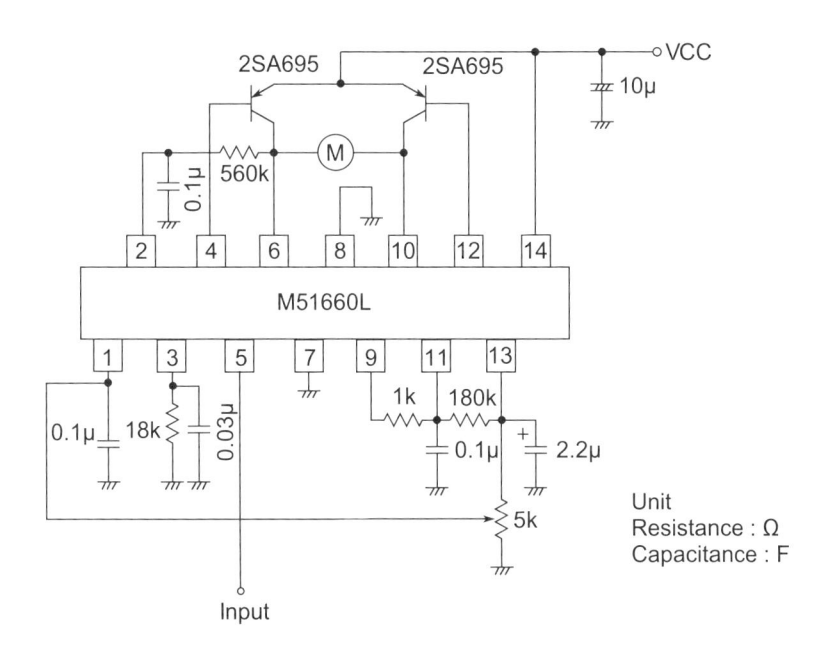

　　　　　　　圖 11-6　　使用 M51660 的 Servo 電路

---

[2] http://www.alldatasheet.com/datasheet-pdf/pdf/875/MITSUBISHI/M51660L.html

11-5　硬體實作

Analog Discovery 和 MG90S 之間的連線如圖 11-7 所示。

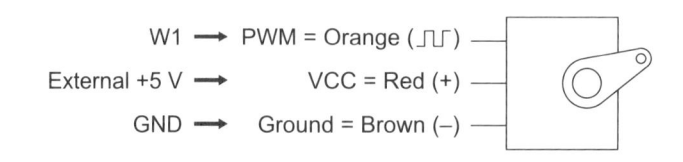

```
W1        →   PWM = Orange (⎍)
External +5 V →   VCC = Red (+)
GND       →   Ground = Brown (−)
```

◎ 圖 11-7 ◎　　　Analog Discovery 和 MG90S 之間的連線

其中 +5 V 電源，因超過 Module 50 mA 的限制，必須外接電源。又因外接電源的 GND 必須與 Analog Discovery 的 GND 連接在一起，為了連接上的方便，使用麵包板來做接線板之用。如實體圖 11-8 所示。

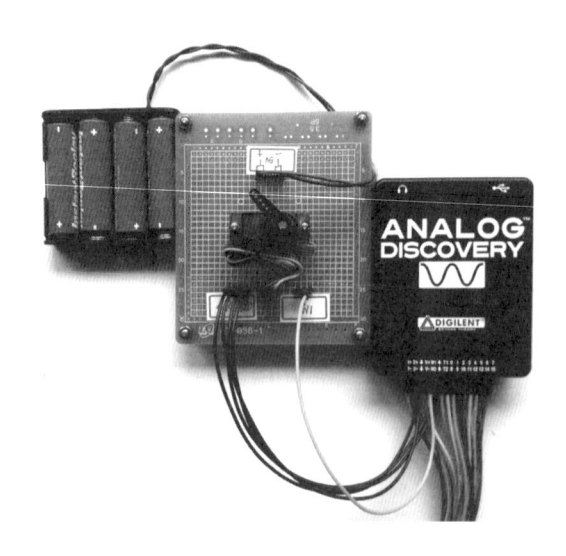

◎ 圖 11-8 ◎　　　Analog Discovery 和 MG90S 之間的實體連線

第十一章　Servo System 與 Control　123

　　由於 +5 V 電源由外部所提供，Analog Discovery 對 MG90S 只提供控制訊號而已，由 Analog 部份的 W1 負責。

　　W1 的設定：首選 Basic，方波。

　　然後：　Frequency 為 50 Hz (20 ms)，
　　　　　　Amplitude 為 2 V，
　　　　　　Offset 為 2 V，
　　　　　　Symmetry 為 5% ~ 10% 之間 (1.0 ms ~ 2.0 ms)，

全部如圖 11-9 所示。

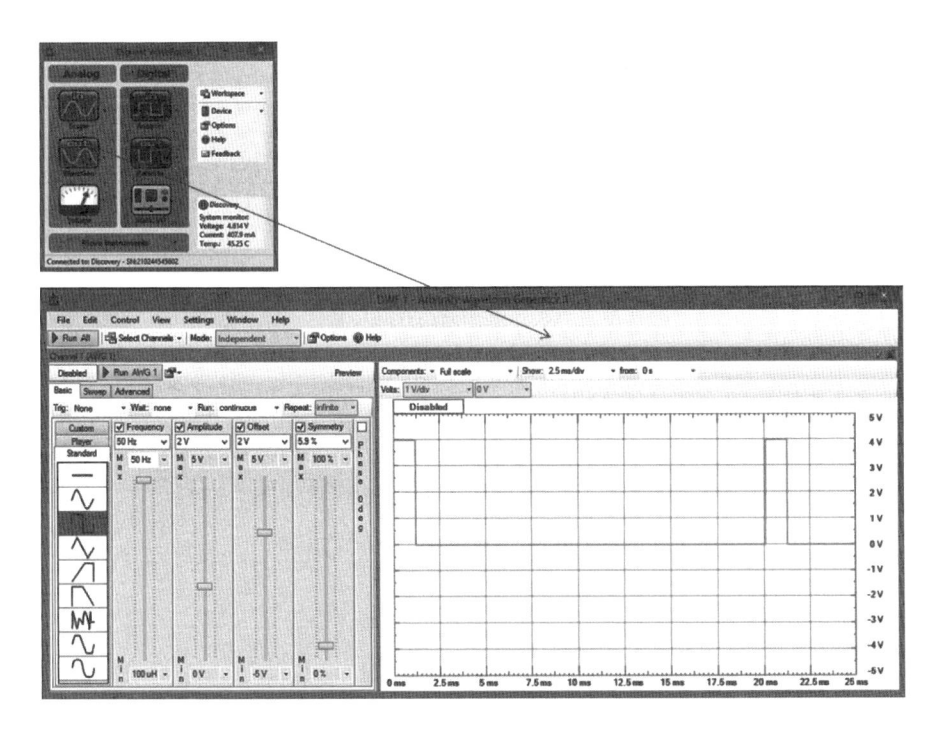

◎ 圖 11-9 ◎　　　W1 對控制 Servo 的設定

　　上下移動 Symmetry 控制鈕，當可觀察到 Servo 的控制桿在 0° 到 180° 間旋轉。

11-6　課外練習

1. 試用 WaveForms 的 Digital Out 來產生相當於 Analog Out 圖 11-7 W1 的 Servo 控制訊號。而且能夠在 0 度、45 度、90 度、135 度和 180 度上停留 (可以加接 Logic 控制電路)。

2. 如何利用 IC NE555 來達成 Servo 的旋轉控制？請用 Ltspice 來做 Simulation 並證實之。

第十二章　Altera/Quartus DE2-115 FPGA 的電路合成

FPGA 是合成邏輯電路的主要元件，1990 年代之後，已漸漸從替代成為取代 SSI 和 MSI 的 IC 邏輯電路。前面數章裡的邏輯電路模擬測試，使用 VHDL 軟體 Modelsim。FPGA 的合成邏輯電路軟體，本章將使用由硬體供應商 Altera 所提供的 Quatus II，實作平台使用的是 Terasic/Alera DE2-115。還有 Xilinx 的 CoolRunner-II CPLD Starter Kit，請參考附錄 G。圖 12-1 是 DE2-115 的相片，上面有許多 Slide-switches、Leds、SSD 和 LCD 等輔助測試用的裝置。為了更能觀察邏輯電路間的 Timing 關係，本章將使用 Analog Discovery Module 中的 Logic Analyzer 和 Logic Pattern Generator 來替代。

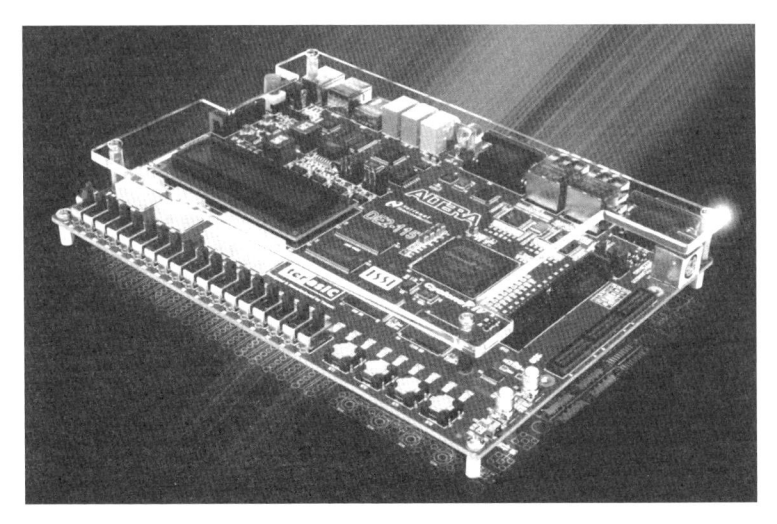

圖 12-1　實作平台 DE2-115 全貌

12-1　DE2-115 的 Expansion Header

為了配合 Analog Discovery Module 以便觀測邏輯電路不同訊號間的 Timing 關係，與 DE2-115 最容易的連接點，當為 Expansion Header JP4 和 JP5。

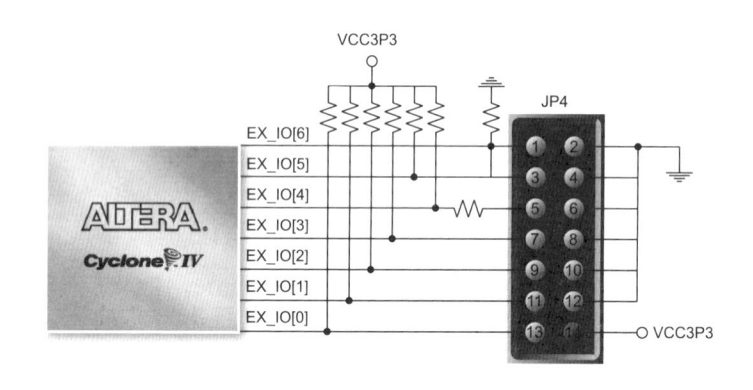

◎ 圖 12-2 ◎　　JP4 Expansion Header 與 FPGA 間的關係

JP4 Expansion Header，如圖 12-2 所示。它的訊號名稱和在 FPGA 的接腳編號，如圖 12-3 所示。

Signal Name	FPGA Pin No.	Description	I/O Standard
EX_IO[0]	PIN_J10	Extended IO[0]	3.3V
EX_IO[1]	PIN_J14	Extended IO[1]	3.3V
EX_IO[2]	PIN_H13	Extended IO[2]	3.3V
EX_IO[3]	PIN_H14	Extended IO[3]	3.3V
EX_IO[4]	PIN_F14	Extended IO[4]	3.3V
EX_IO[5]	PIN_E10	Extended IO[5]	3.3V
EX_IO[6]	PIN_D9	Extended IO[6]	3.3V

◎ 圖 12-3 ◎　　JP4 Expansion Header 在 FPGA 的訊號名稱和接腳編號

第十二章　Altera/Quartus DE2-115 FPGA 的電路合成　127

　　JP5 Expansion Header，如圖 12-4 所示。它的訊號名稱和在 FPGA 的接腳編號，如圖 12-6 所示。其中的 5 V 和 3.3 V 電源，可以提供的最大電流，如圖 12-5 所示。

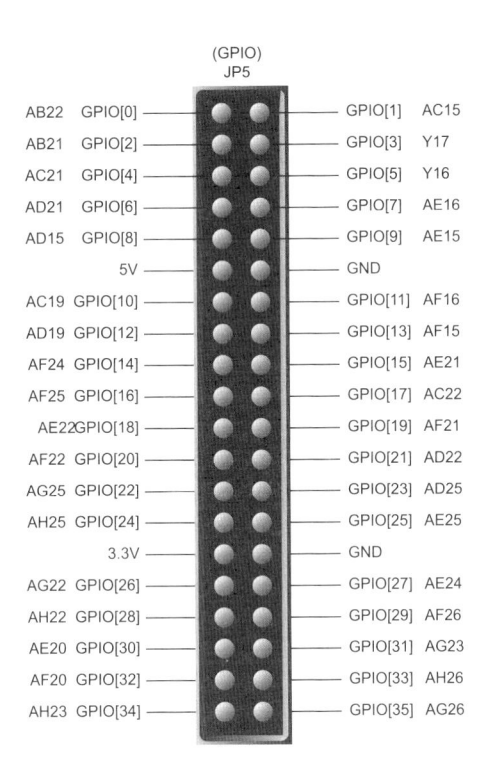

```
                    (GPIO)
                     JP5
    AB22  GPIO[0]  ─●  ●─  GPIO[1]   AC15
    AB21  GPIO[2]  ─●  ●─  GPIO[3]   Y17
    AC21  GPIO[4]  ─●  ●─  GPIO[5]   Y16
    AD21  GPIO[6]  ─●  ●─  GPIO[7]   AE16
    AD15  GPIO[8]  ─●  ●─  GPIO[9]   AE15
           5V      ─●  ●─   GND
    AC19  GPIO[10] ─●  ●─  GPIO[11]  AF16
    AD19  GPIO[12] ─●  ●─  GPIO[13]  AF15
    AF24  GPIO[14] ─●  ●─  GPIO[15]  AE21
    AF25  GPIO[16] ─●  ●─  GPIO[17]  AC22
    AE22  GPIO[18] ─●  ●─  GPIO[19]  AF21
    AF22  GPIO[20] ─●  ●─  GPIO[21]  AD22
    AG25  GPIO[22] ─●  ●─  GPIO[23]  AD25
    AH25  GPIO[24] ─●  ●─  GPIO[25]  AE25
          3.3V     ─●  ●─   GND
    AG22  GPIO[26] ─●  ●─  GPIO[27]  AE24
    AH22  GPIO[28] ─●  ●─  GPIO[29]  AF26
    AE20  GPIO[30] ─●  ●─  GPIO[31]  AG23
    AF20  GPIO[32] ─●  ●─  GPIO[33]  AH26
    AH23  GPIO[34] ─●  ●─  GPIO[35]  AG26
```

圖 12-4　JP5 Expansion Header 與 FPGA 間的關係

Supplied Voltage	Max. Current Limit
5V	1 A
3.3 V	1.5 A

圖 12-5　5 V 和 3.3 V 電源可以提供的最大電流

Signal Name	FPGA Pin No.	Description	I/O Standard
GPIO[0]	PIN_AB22	GPIO Connection DATA[0]	Depending on JP6
GPIO[1]	PIN_AC15	GPIO Connection DATA[1]	Depending on JP6
GPIO[2]	PIN_AB21	GPIO Connection DATA[2]	Depending on JP6
GPIO[3]	PIN_Y17	GPIO Connection DATA[3]	Depending on JP6
GPIO[4]	PIN_AC21	GPIO Connection DATA[4]	Depending on JP6
GPIO[5]	PIN_Y16	GPIO Connection DATA[5]	Depending on JP6
GPIO[6]	PIN_AD21	GPIO Connection DATA[6]	Depending on JP6
GPIO[7]	PIN_AE16	GPIO Connection DATA[7]	Depending on JP6
GPIO[8]	PIN_AD15	GPIO Connection DATA[8]	Depending on JP6
GPIO[9]	PIN_AE15	GPIO Connection DATA[9]	Depending on JP6
GPIO[10]	PIN_AC19	GPIO Connection DATA[10]	Depending on JP6
GPIO[11]	PIN_AF16	GPIO Connection DATA[11]	Depending on JP6
GPIO[12]	PIN_AD19	GPIO Connection DATA[12]	Depending on JP6
GPIO[13]	PIN_AF15	GPIO Connection DATA[13]	Depending on JP6
GPIO[14]	PIN_AF24	GPIO Connection DATA[14]	Depending on JP6
GPIO[15]	PIN_AE21	GPIO Connection DATA[15]	Depending on JP6
GPIO[16]	PIN_AF25	GPIO Connection DATA[16]	Depending on JP6
GPIO[17]	PIN_AC22	GPIO Connection DATA[17]	Depending on JP6
GPIO[18]	PIN_AE22	GPIO Connection DATA[18]	Depending on JP6
GPIO[19]	PIN_AF21	GPIO Connection DATA[19]	Depending on JP6
GPIO[20]	PIN_AF22	GPIO Connection DATA[20]	Depending on JP6
GPIO[21]	PIN_AD22	GPIO Connection DATA[21]	Depending on JP6
GPIO[22]	PIN_AG25	GPIO Connection DATA[22]	Depending on JP6
GPIO[23]	PIN_AD25	GPIO Connection DATA[23]	Depending on JP6
GPIO[24]	PIN_AH25	GPIO Connection DATA[24]	Depending on JP6
GPIO[25]	PIN_AE25	GPIO Connection DATA[25]	Depending on JP6
GPIO[26]	PIN_AG22	GPIO Connection DATA[26]	Depending on JP6
GPIO[27]	PIN_AE24	GPIO Connection DATA[27]	Depending on JP6
GPIO[28]	PIN_AH22	GPIO Connection DATA[28]	Depending on JP6
GPIO[29]	PIN_AF26	GPIO Connection DATA[29]	Depending on JP6
GPIO[30]	PIN_AE20	GPIO Connection DATA[30]	Depending on JP6
GPIO[31]	PIN_AG23	GPIO Connection DATA[31]	Depending on JP6
GPIO[32]	PIN_AF20	GPIO Connection DATA[32]	Depending on JP6
GPIO[33]	PIN_AH26	GPIO Connection DATA[33]	Depending on JP6
GPIO[34]	PIN_AH23	GPIO Connection DATA[34]	Depending on JP6
GPIO[35]	PIN_AG26	GPIO Connection DATA[35]	Depending on JP6

圖 12-6　JP5 Expansion Header 在 FPGA 的訊號名稱和接腳編號

JP5 接腳的輸入或輸出電位，可以經由改變 JP6 的 Jumper 的位置而改變，如圖 12-7 所示。

JP6 Jumper Settings	Supplied Voltage to VCCIO4	IO Voltage of Expansion Headers (JP5)
Short Pins 1 and 2	1.5 V	1.5V
Short Pins 3 and 4	1.8 V	1.8 V
Short Pins 5 and 6	2.5 V	2.5 V
Short Pins 7 and 8	3.3 V	3.3 V (Default)

圖 12-7　JP6 Jumper 的設定與 JP5 接腳的輸入或輸出電位的關係

12-2　Synthesis 軟體 Quartus II / VHDL 的介紹

　　Altera 的 Quartus II 軟體，有 Schematic Design、VHDL Design 和 Verilog Design 等三種。為配合先前使用的 Modelsim/VHDL Simulator，本章所使用的是 VHDL Design。不同於 Design 電路，Synthesis 電路比較機械化。它不像 VHDL 電路的設計，而必須按照一定的程序進行，少有使用者可以變動的地方。除了顯得繁雜，但不困難。當電路的 VHDL 檔通過 TestBench 檔的 Simulation 後，該 VHDL 檔就可以用 FPGA 來做 Synthesis。合成電路的程序如下：

1. Starting a New Project
   決定 Project 的名稱和位置，及 VHDL 電路的檔名，使用 FPGA 的編號。
2. First Compilation the Design
   第一次對 Project 的 VHDL 電路的檔的編輯。
3. Pin Assignment
   決定電路輸入和輸出端在 FPGA 接腳上的位置。

4. Second Compilation the Design

   第二次對 Project 的 VHDL 電路的檔 FPGA 接腳的編輯。

5. Programming and Configuring the FPGA Device

   把完成的電路 Bit Pattern 直接加入到 FPGA 上。

6. Test the Designed Circuit

   使用 Analog Discovery 的 Logic Pattern Generator 和 Logic Analyzer 來測試所設計完成的電路是否工作正常。

7. Active Serial Mode Programming

   當 Power off 後，FPGA 將失去程序 5 所有加入的 Bit Pattern，如果要能在 Power on 之後，FPGA 立刻動作，必須另加 Configuration Device，DE2-115 的 Configuration Device 是編號為 EPCS64 的 EEPROM。當 Power off 之後，EEPROM 還能保持內有 Data Pattern 一段很長的時間，當 Power on 的時候，可以很快地存回 FPGA。所以叫做 Active Serial Mode Programming。

8. Third Compilation the Design

   選用了 Configuration Device 之後，也需要做一次編輯。

9. Programming and Configuring the FPGA Device

   把完成的電路 Bit Pattern 直接加入到 EEPROM 上。

10. Test the Designed Circuit

    使用與程序 6 的裝置設定，來測試所設計完成的電路是否工作正常。

從以上 10 點，使人感覺到整個 Synthesis 過程的繁雜，但從經驗上得知，只要一步步小心地做，多做幾個電路，熟悉之後，也就無所謂繁雜了。

12-3　Synthesis 電路的一個例子

　　這個電路來自第六章的 counter.vhd。當 Quartus II 啟動之後，點擊 "**New Project Wizard**"，如圖 S1-1。

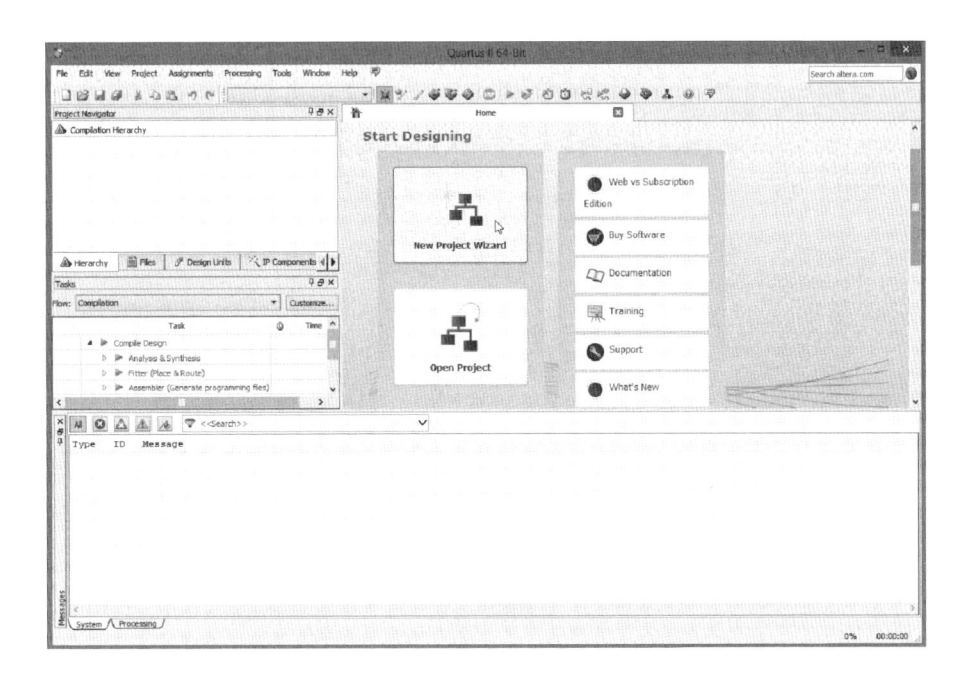

圖 S1-1　　進入 New Project Wizard

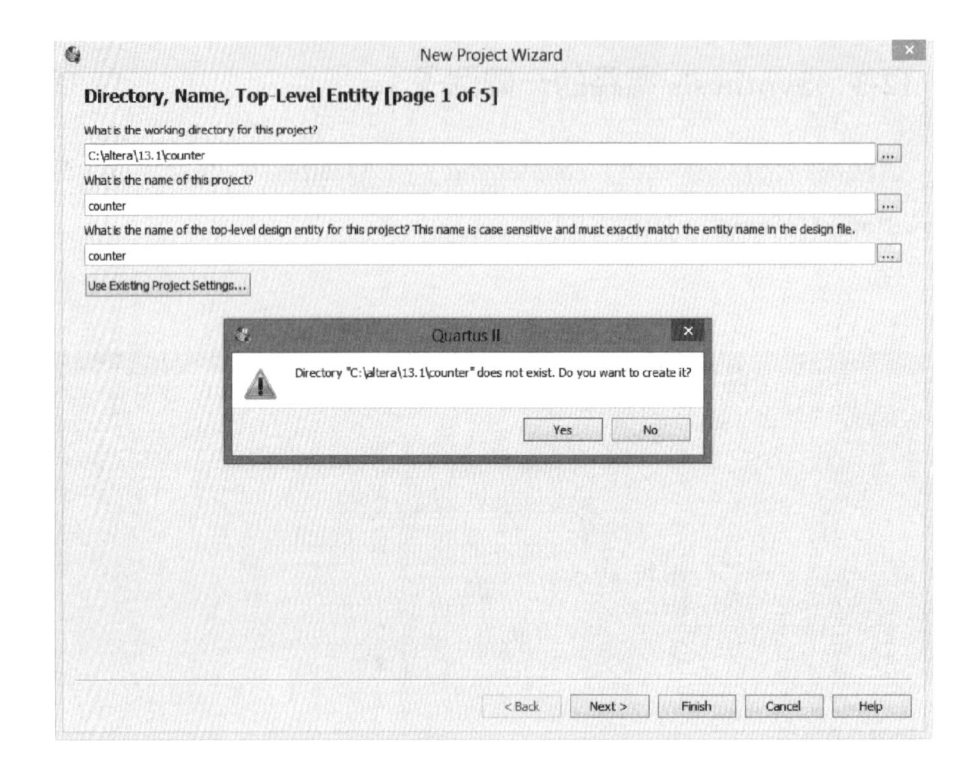

◦ 圖 S1-2 ◦　　　設定檔案夾，和最高層設計 Entity 的名稱

　　圖 S1-2 之後，可以跳過 [page 2 of 5]。來到圖 S1-3 的 Family & Device Setting。

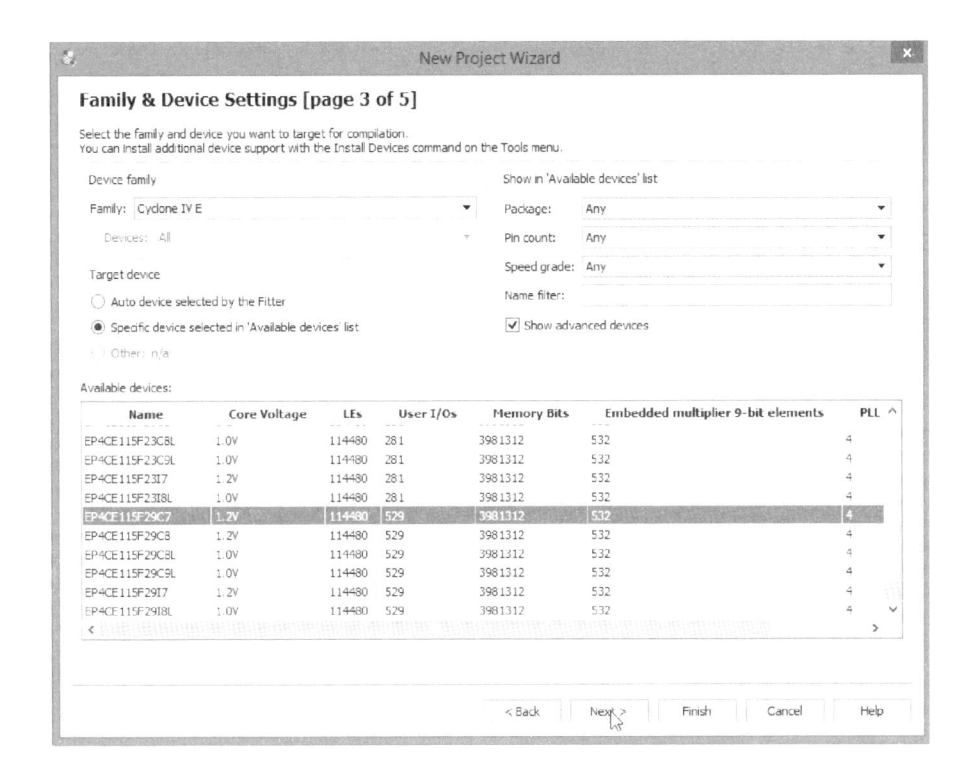

第十二章　Altera/Quartus DE2-115 FPGA 的電路合成　133

圖 S1-3：　　　FPGA 群屬與編號的選用

　　DE2-115 所用的 FPGA 屬於 Cyclone IV E Family 的 EP4CE115F29C7。之後又可以跳過 [page 4 of 5]。最後來到了圖 S1-4 的**結論** (Summary)。

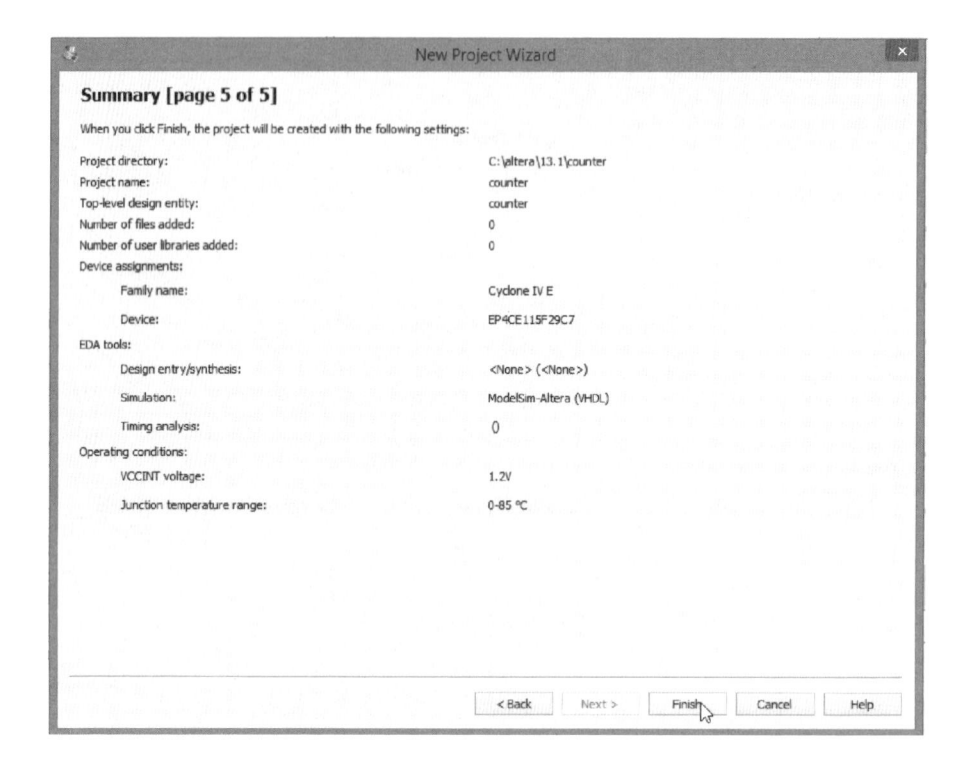

```
New Project Wizard
Summary [page 5 of 5]
When you click Finish, the project will be created with the following settings:

Project directory:                              C:\altera\13.1\counter
Project name:                                   counter
Top-level design entity:                        counter
Number of files added:                          0
Number of user libraries added:                 0
Device assignments:
    Family name:                                Cyclone IV E
    Device:                                     EP4CE115F29C7
EDA tools:
    Design entry/synthesis:                     <None> (<None>)
    Simulation:                                 ModelSim-Altera (VHDL)
    Timing analysis:                            0
Operating conditions:
    VCCINT voltage:                             1.2V
    Junction temperature range:                 0-85 °C

                        < Back    Next >    Finish    Cancel    Help
```

圖 S1-4　　　New Project Wizard 的結論

在這個 S1 的階段裡，counter 檔案夾的產物如圖 S1-5 所示。

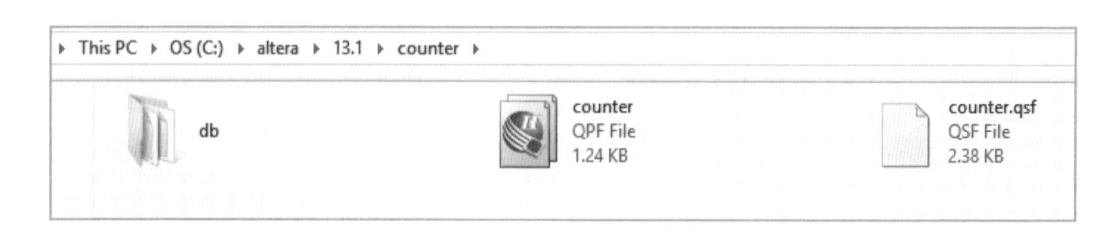

```
This PC ▶ OS (C:) ▶ altera ▶ 13.1 ▶ counter ▶

    db                      counter              counter.qsf
                            QPF File             QSF File
                            1.24 KB              2.38 KB
```

圖 S1-5　　　S1 的階段裡，counter 檔案夾的產物

第十二章　Altera/Quartus DE2-115 FPGA 的電路合成　135

S2 階段從選用 File > New 開始，如圖 S2-1，這裡應選用 VHDL File。

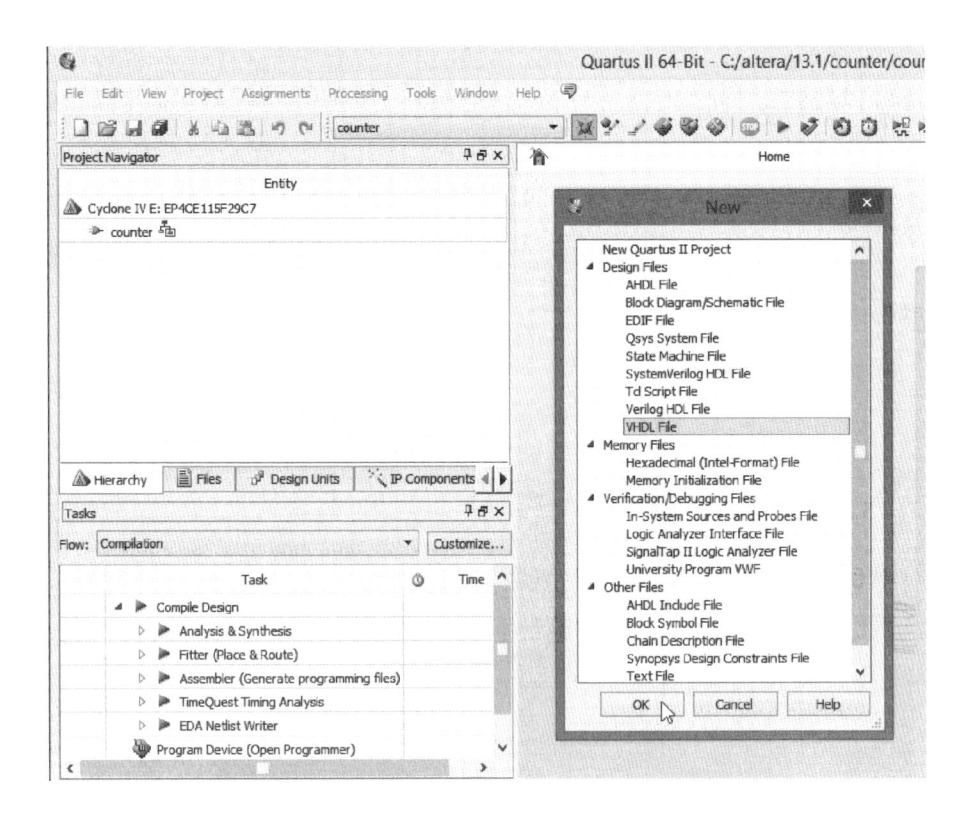

▓ 圖 S2-1 ▓　　　　Project 中設計檔案的類別

接下來是選用 File > Save As，如圖 S2-2 (Save 的是 File 的名稱)。再把第六章的 Counter.vhd，Copy 到空白的 File 上，如圖 S2-3 所示 (請記得 Copy 完了後要 Save)。

圖 S2-4 是要將所設計的檔案，加到 Project 中。選用 Assignment > Setting。

圖 S2-5 顯示所設計的檔案 "counter.vhd" 已經加入到 **Project** 中。

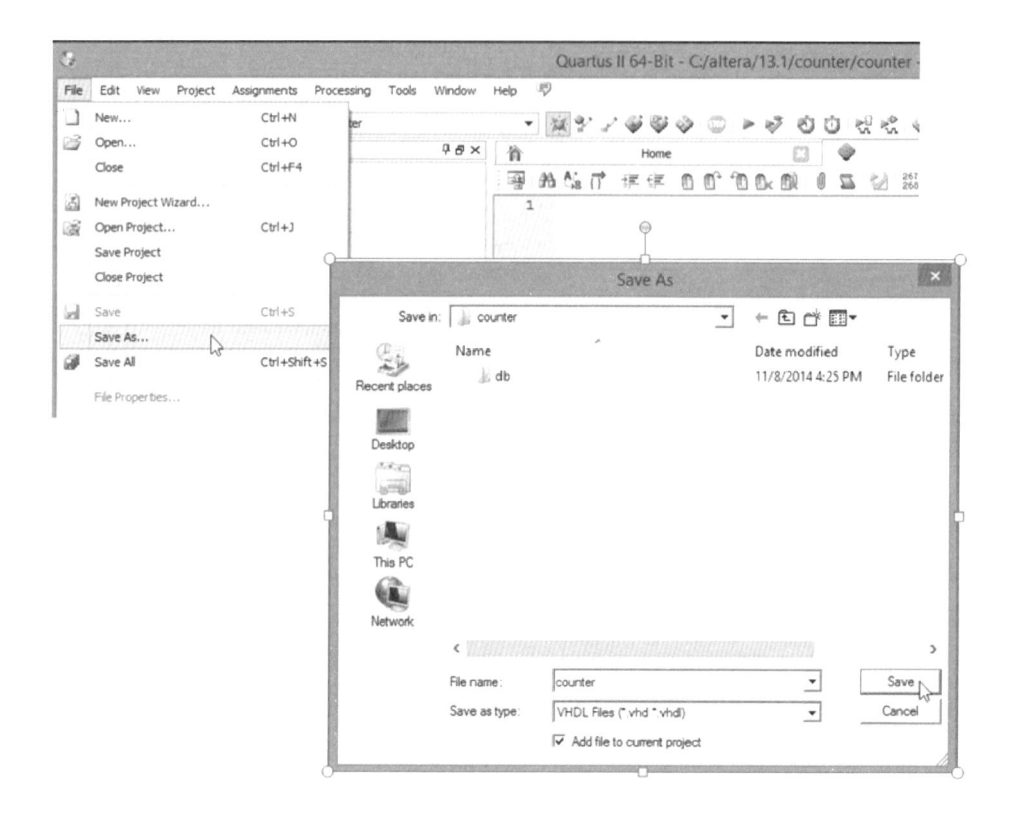

◉ 圖 S2-2 ◉　　　Save 的是設計檔案的名稱

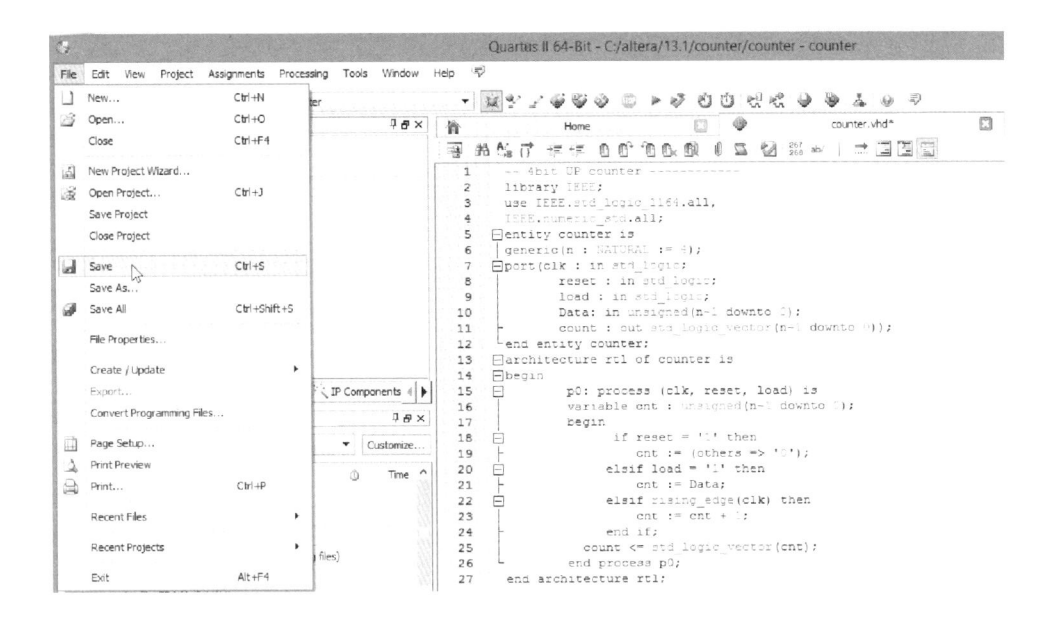

第十二章　Altera/Quartus DE2-115 FPGA 的電路合成

圖 S2-3　把第六章的 Counter.vhd，Copy 到空白的 File 上

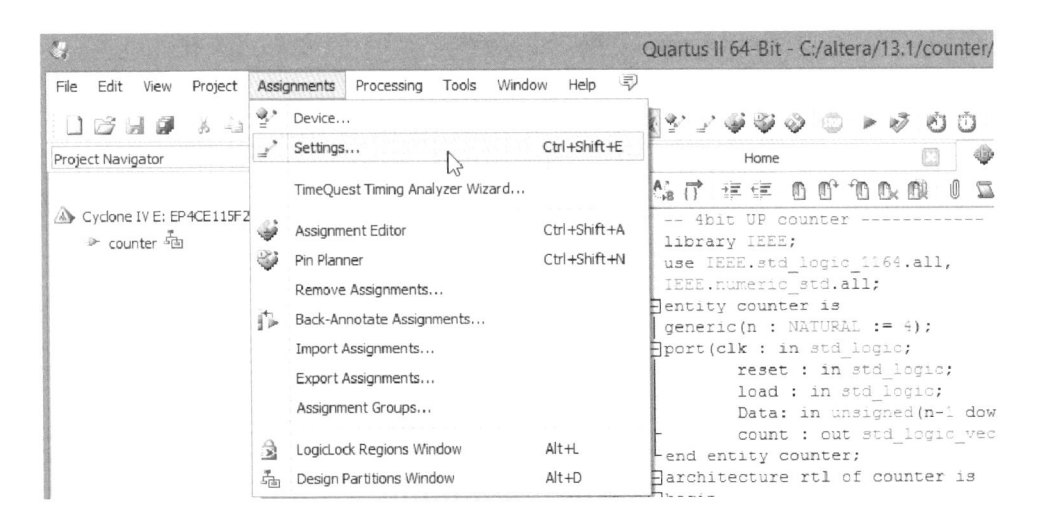

圖 S2-4　設定將所設計的檔案加到 Project 中

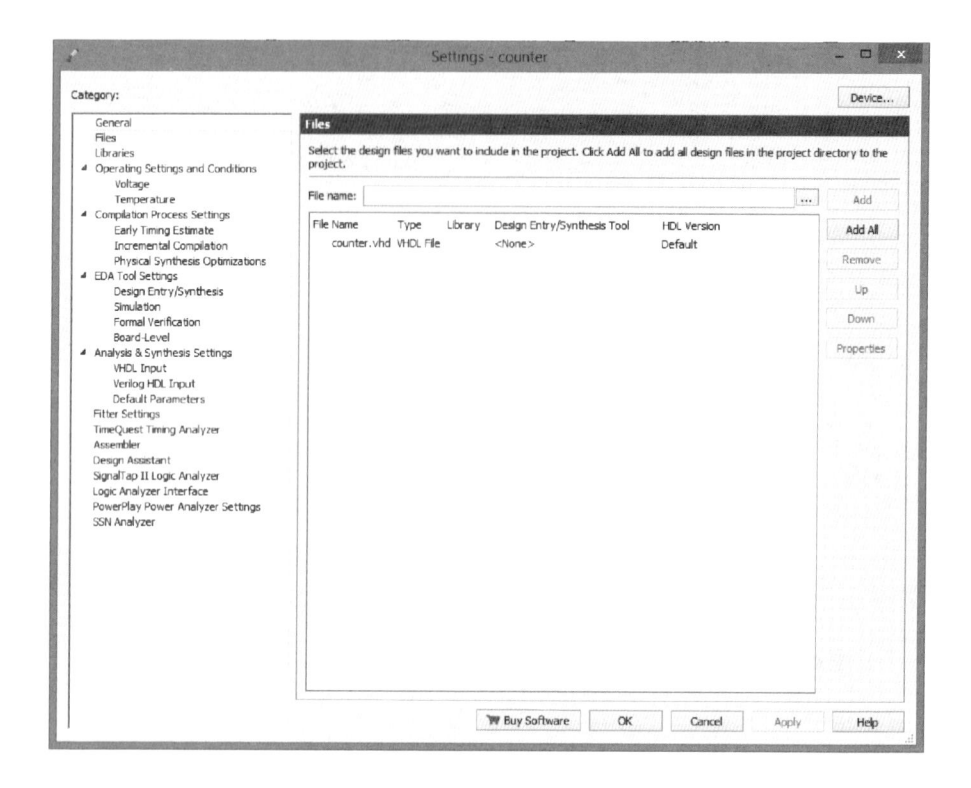

◎ 圖 S2-5 ◎　　　證實 Design File "counter.vhd" 已在 Project 中

在這個 S2 的階段裡，counter 檔案夾的產物如圖 S2-6 所示。

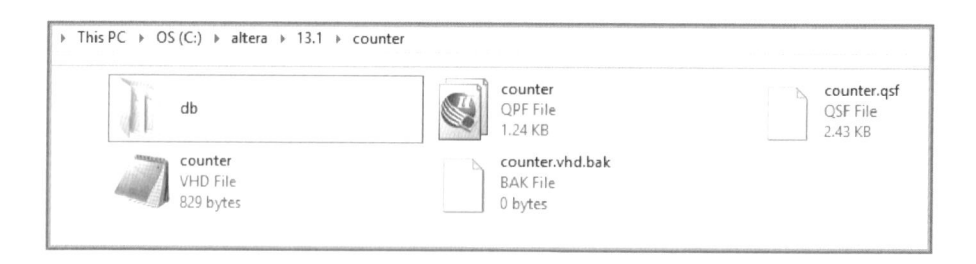

◎ 圖 S2-6 ◎　　　S2 的階段裡，counter 檔案夾的產物

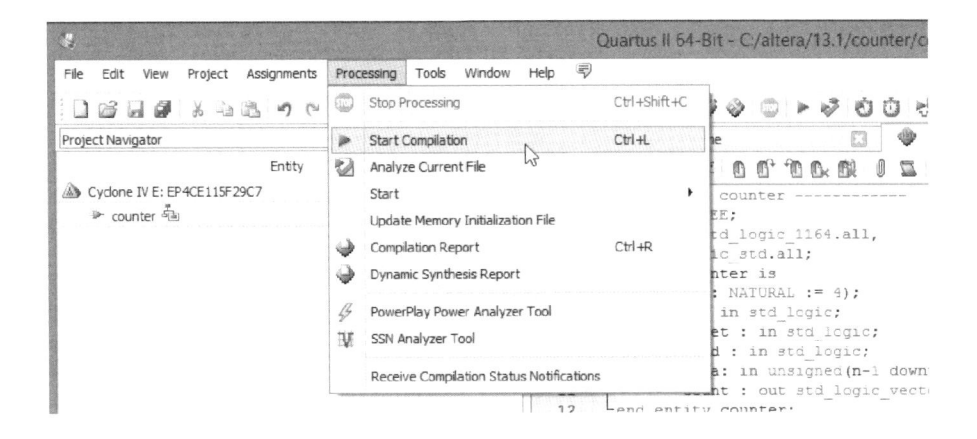

第十二章　Altera/Quartus DE2-115 FPGA 的電路合成　139

圖 S2-7: 第一次**編輯** (Compilation) 的開始

圖 S2-7　　第一次編輯 (Compilation) 的開始

140　iLAB Digital 數位電路設計、模擬測試與硬體除錯

圖 S2-8: Compilation 成功

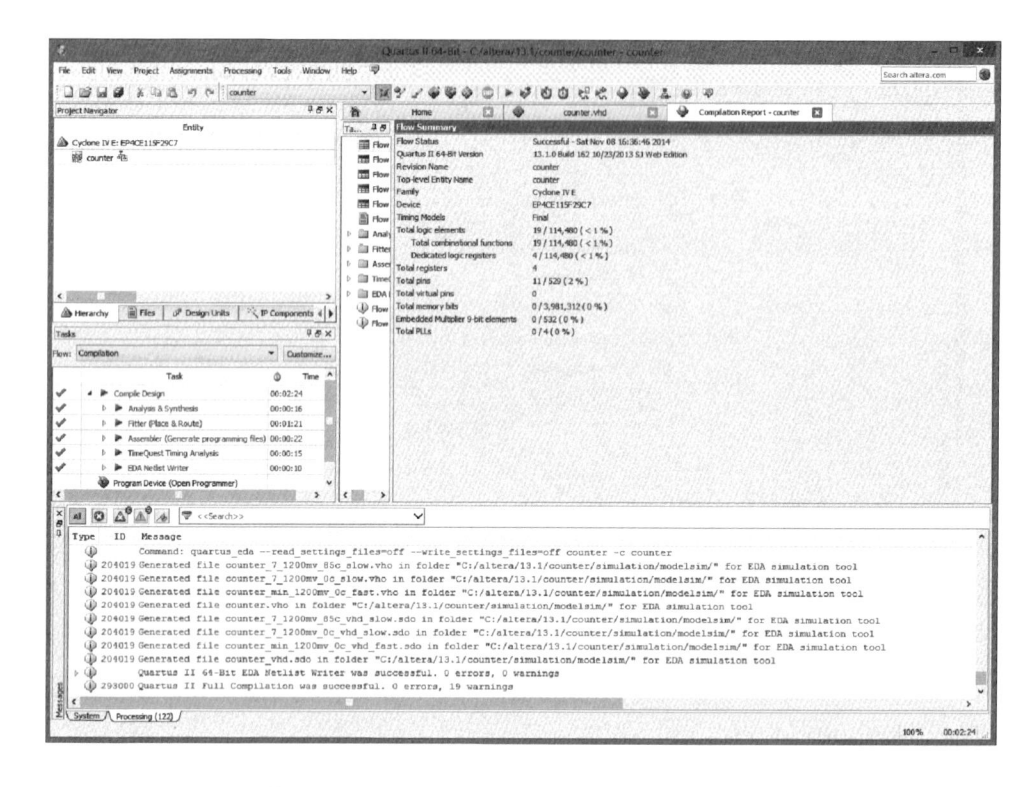

◦⬚ 圖 S2-8 ⬚◦　　　編輯 (Compilation) 成功

　　接下來的是第三階段 S3，決定電路輸入和輸出端在 FPGA 接腳上的位置。

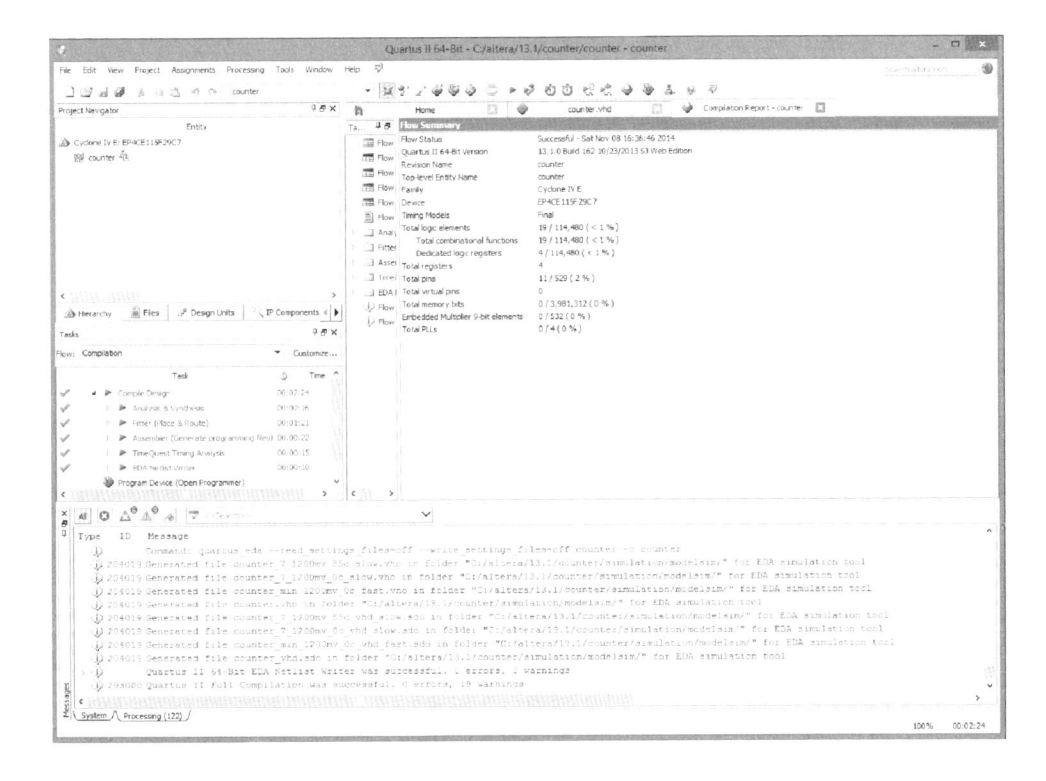

第十二章　Altera/Quartus DE2-115 FPGA 的電路合成

圖 S3-1 選用 Assignments > Assignment Editor。

圖 S3-1　選用 Assignments > Assignment Editor

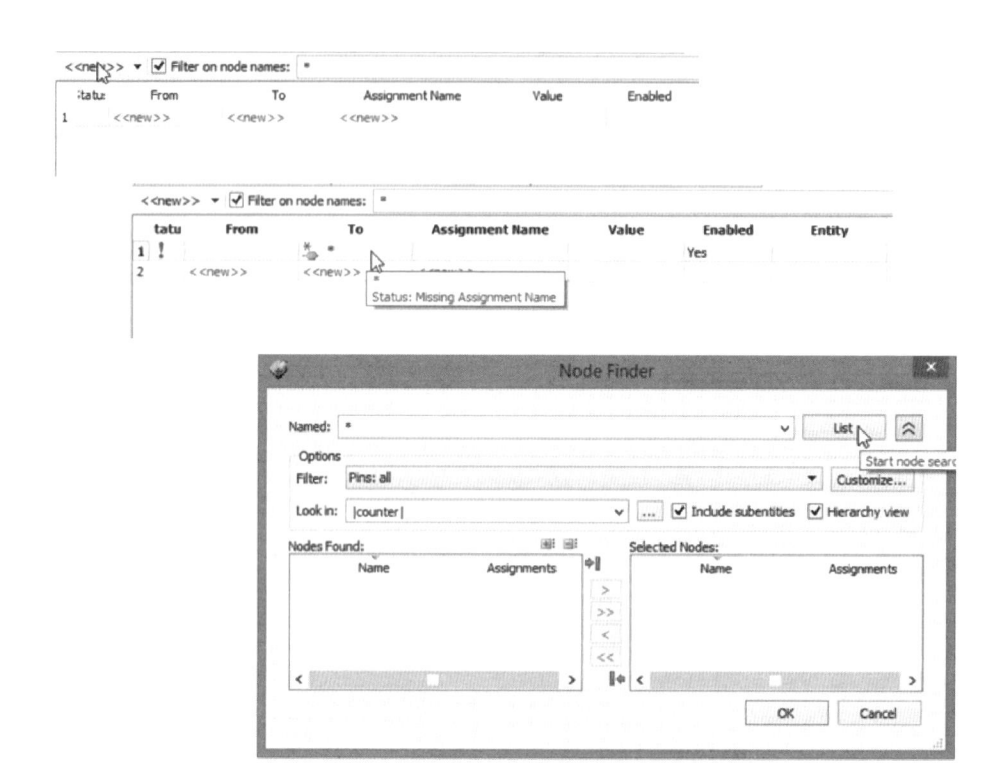

142　iLAB Digital 數位電路設計、模擬測試與硬體除錯

圖 S3-2 Quartus II 一連串的程序，從 Design File 中尋取它的 I/O 接點。

◎ 圖 S3-2　　　Quartus II 一連串的程序，從 Design File 中尋取它的 I/O 接點

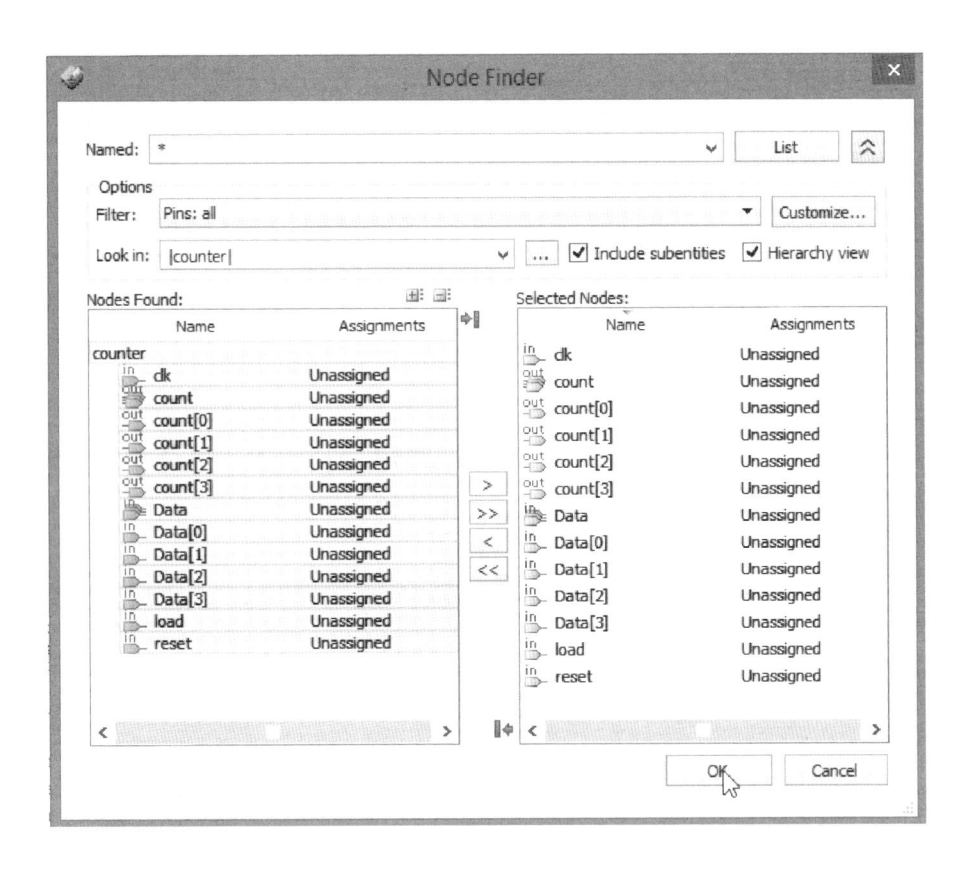

第十二章 Altera/Quartus DE2-115 FPGA 的電路合成

圖 S3-3 Quartus II 找到的電路 I/O 接點。

圖 S3-3　　Quartus II 找到的電路 I/O 接點

圖 S3-4: DE2-115/JP5 FPGA 接腳和 Analog Discovery 間的關係。

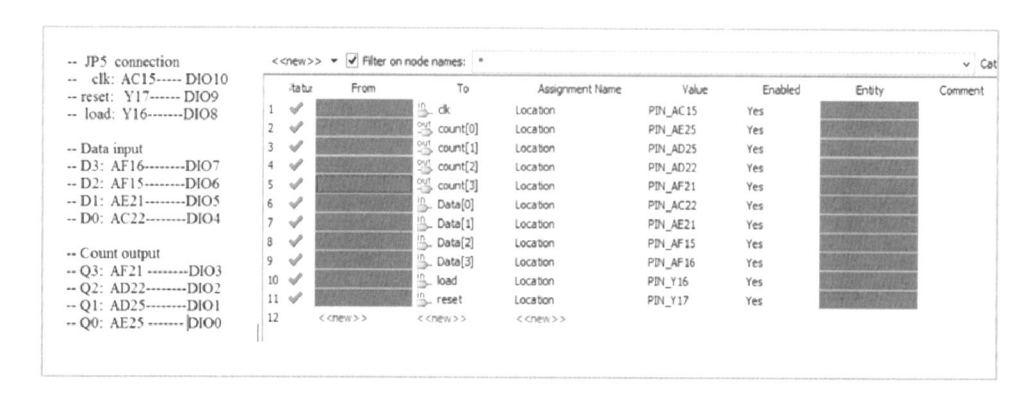

◎ 圖 S3-4 ◎　　DE2-115/JP5　FPGA 接腳和 Analog Discovery 間的關係

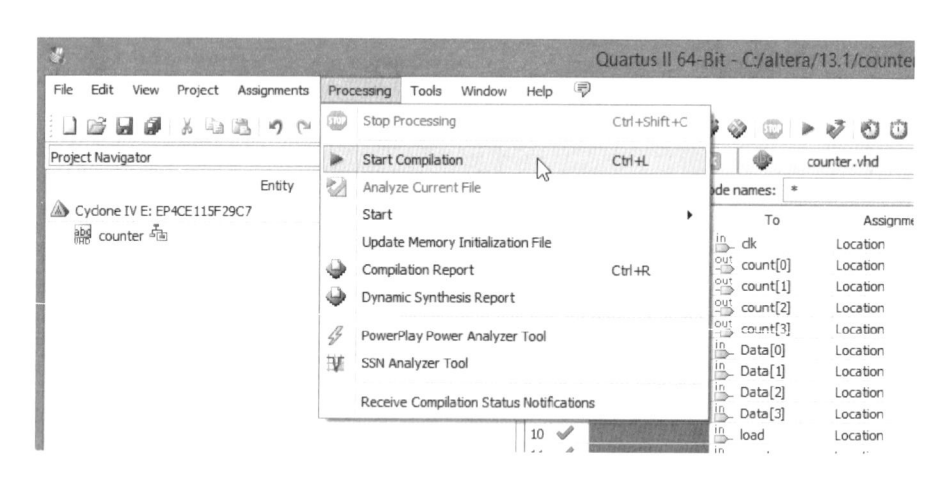

◎ 圖 S3-5 ◎　　第二次編輯 (Compilation) 的開始

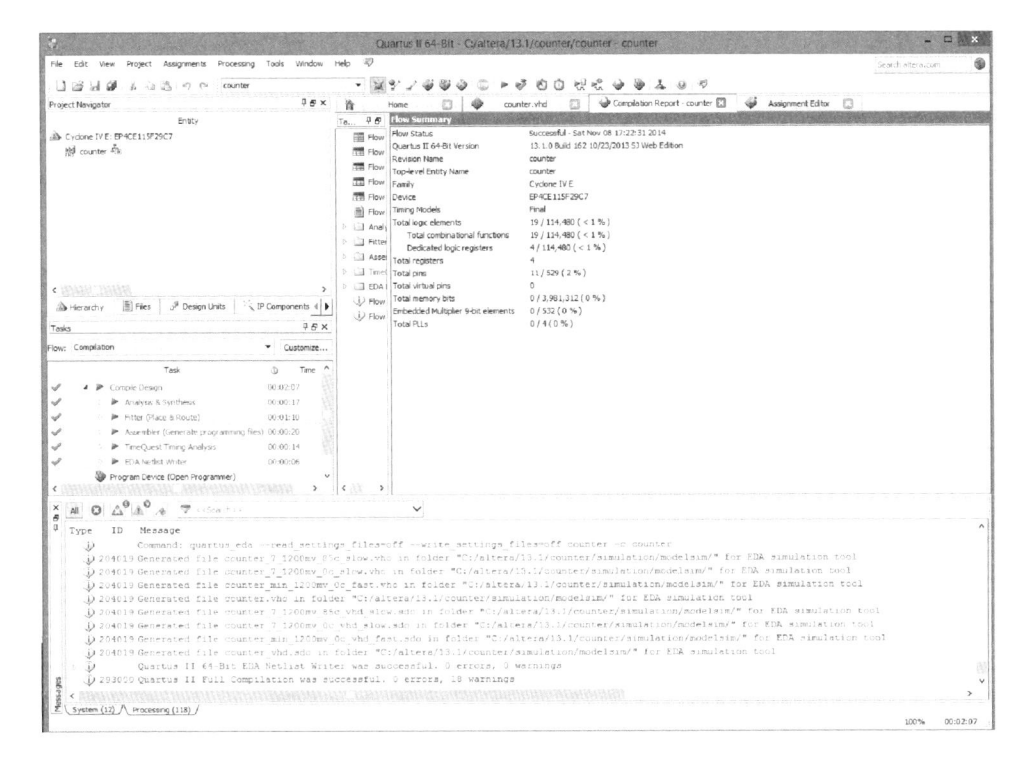

第十二章　Altera/Quartus DE2-115 FPGA 的電路合成

◦ 圖 S3-6 ◦　　　編輯 (Compilation) 成功

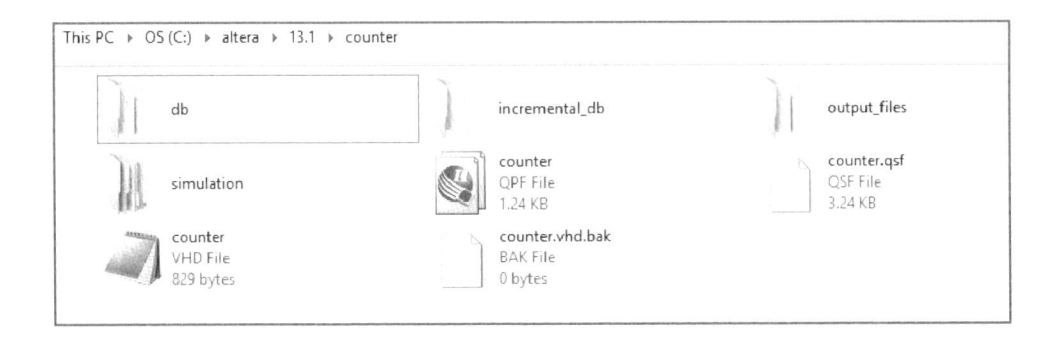

◦ 圖 S3-7 ◦　　　S3 的階段產物 counter.sof 存在於 "output_files" 檔案夾中

　　Counter.sof 可以用來 Program FPGA，但是當 FPGA 被 Power Off，Program 立刻消失無蹤。如果事先存入 EEPROM/ EPCS64，讓它在

Power On 時立刻再存入 FPGA。

圖 S4-1 為選取 EEPROM 的開始，**Assignment > Device**。

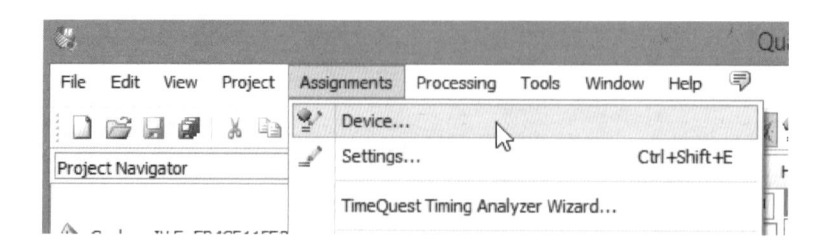

◦ 圖 S4-1 ◦　　選取 EEPROM 的第一步

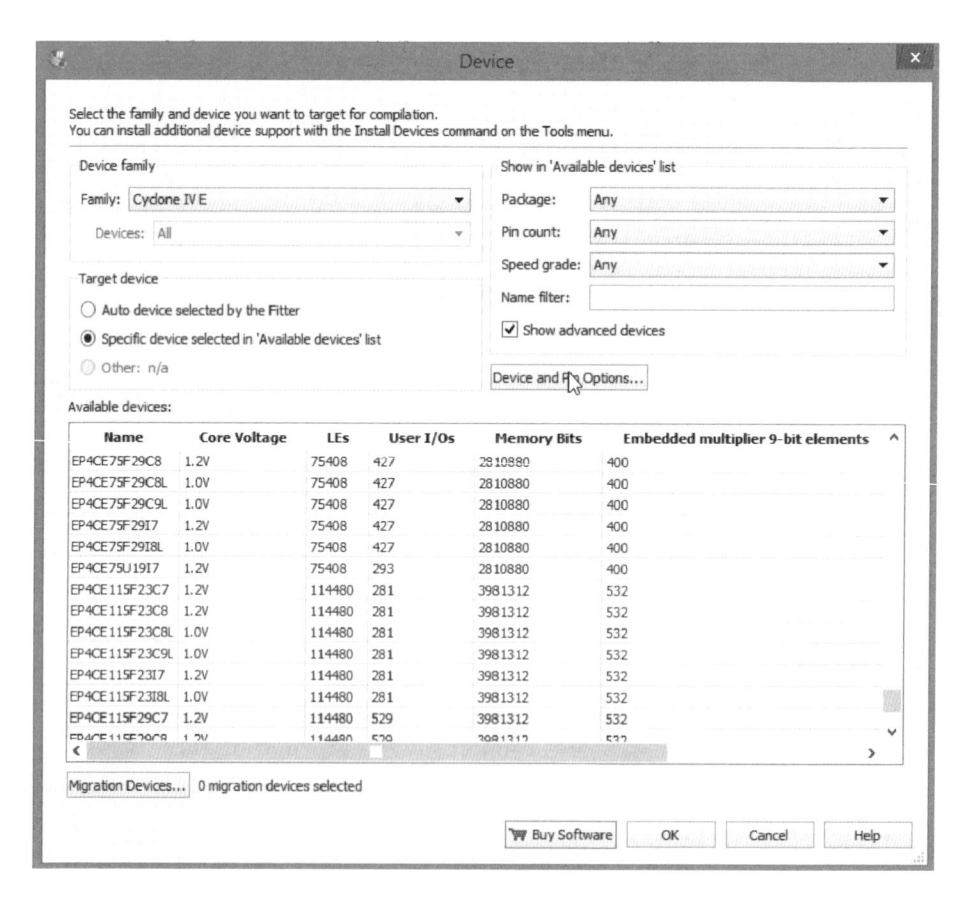

◦ 圖 S4-2 ◦　　選取 EEPROM 的第二步

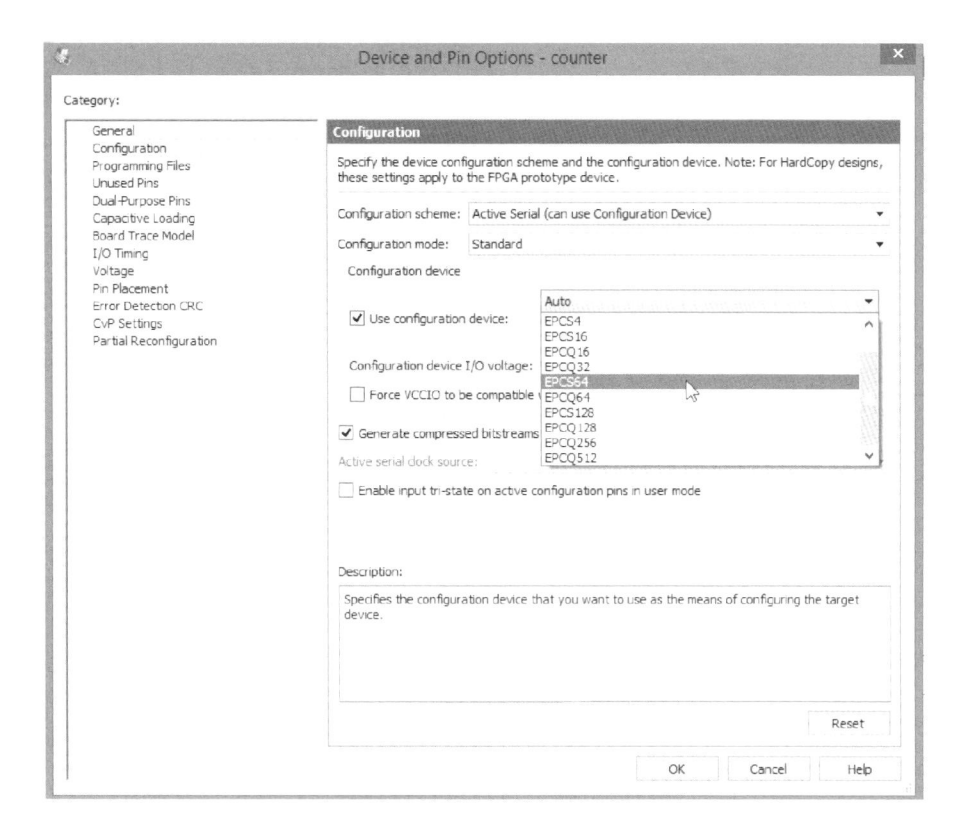

第十二章　Altera/Quartus DE2-115 FPGA 的電路合成　147

圖 S4-3　第三步 DE2-115 應選取 EEPROM/EPCS64

◎ 圖 S4-4　　　　　第四步 OK

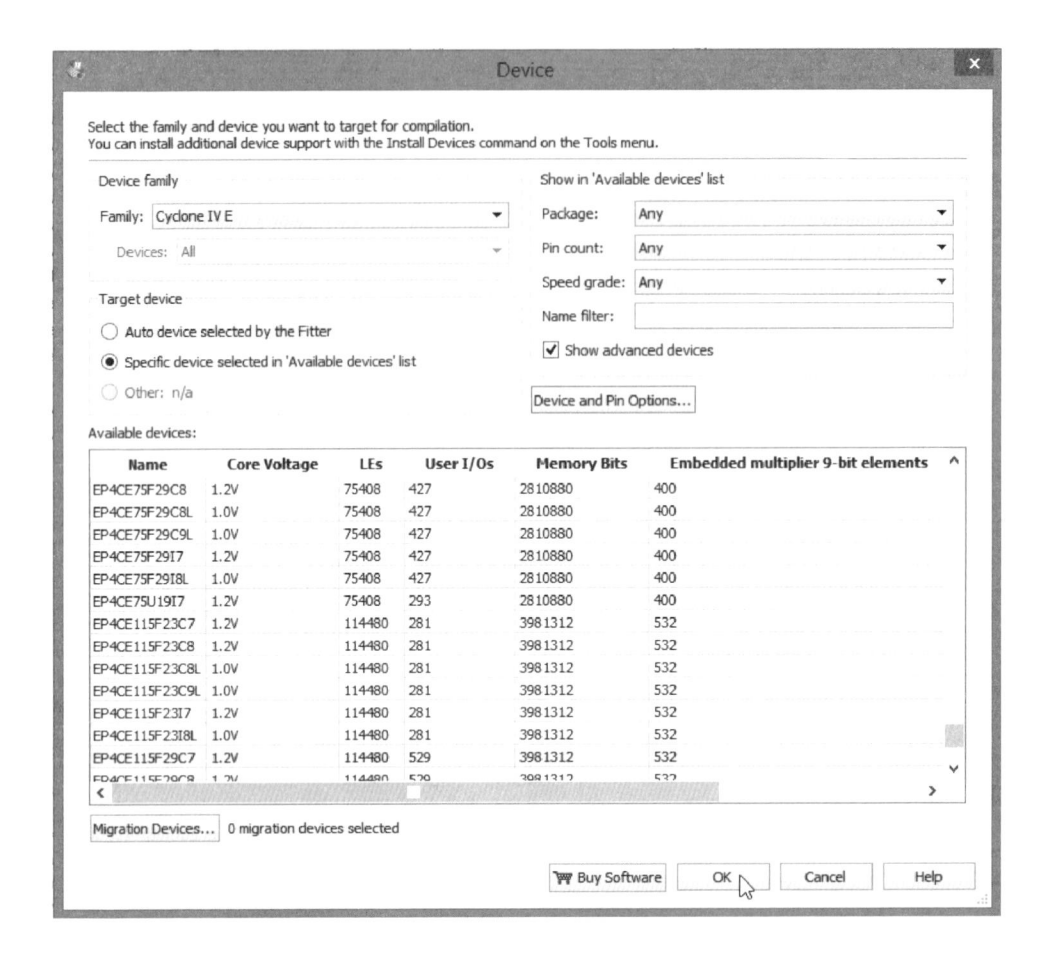

接下來再做一次，也就是第三次的編輯工作，如圖 S4-5 所示。

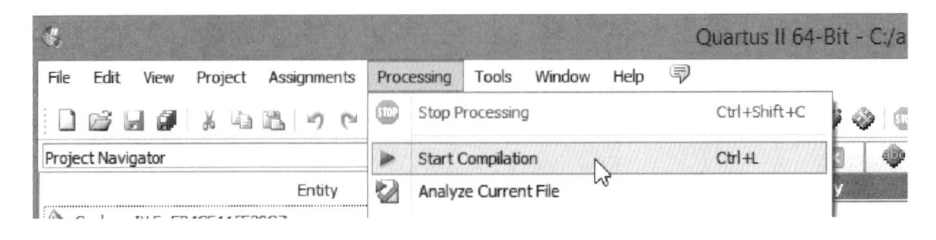

◎ 圖 S4-5　　　　　開始第三次的編輯工作

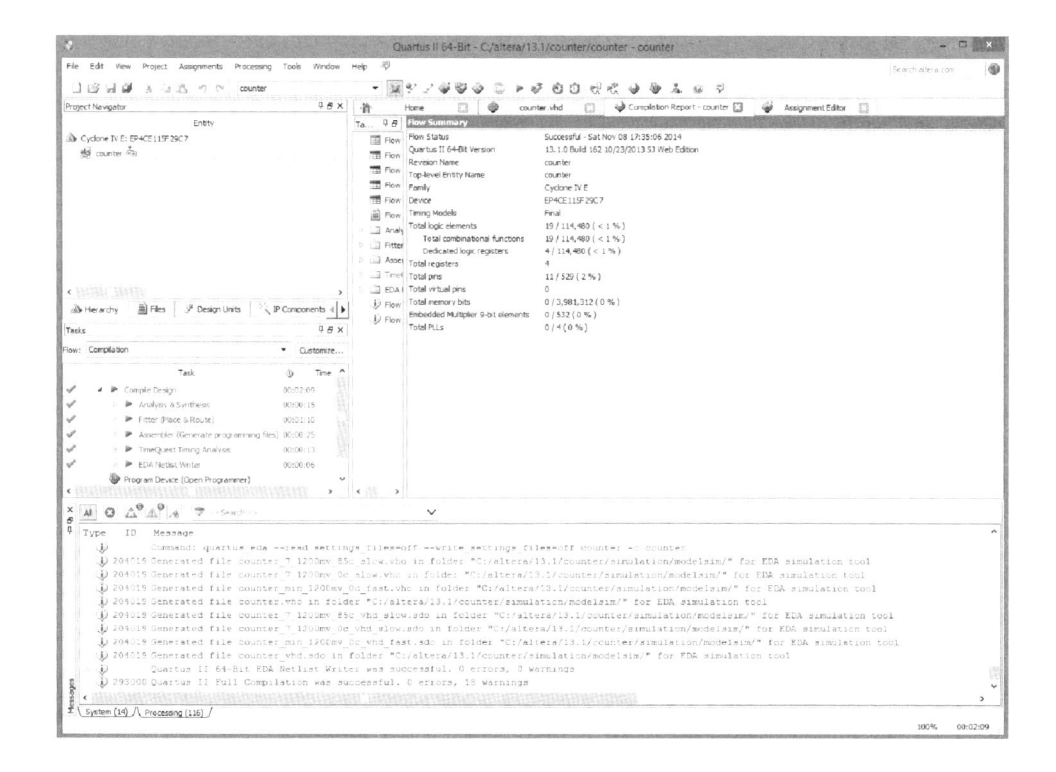

第十二章　Altera/Quartus DE2-115 FPGA 的電路合成

◦ 圖 S4-6 ◦　　　第三次的編輯工作成功

◦ 圖 S4-7 ◦　　　S4 的階段產物 counter.pof 存在於 "output_files" 檔案夾中

圖 S4-8 Counter >> output_files 一覽表，counter.sof 和 counter.pof 全部在內。

Name	Date modified	Type	Size
counter.asm.rpt	11/8/2014 5:34 PM	RPT File	8 KB
counter.done	11/8/2014 5:35 PM	DONE File	1 KB
counter.eda.rpt	11/8/2014 5:35 PM	RPT File	8 KB
counter.fit.rpt	11/8/2014 5:34 PM	RPT File	232 KB
counter.fit.smsg	11/8/2014 5:34 PM	SMSG File	1 KB
counter.fit.summary	11/8/2014 5:34 PM	SUMMARY File	1 KB
counter.flow.rpt	11/8/2014 5:35 PM	RPT File	9 KB
counter.jdi	11/8/2014 5:34 PM	JDI File	1 KB
counter.map.rpt	11/8/2014 5:33 PM	RPT File	22 KB
counter.map.summary	11/8/2014 5:33 PM	SUMMARY File	1 KB
counter.pin	11/8/2014 5:34 PM	PIN File	91 KB
counter.pof	11/8/2014 5:34 PM	POF File	8,193 KB
counter.sof	11/8/2014 5:34 PM	SOF File	3,459 KB
counter.sta.rpt	11/8/2014 5:35 PM	RPT File	143 KB
counter.sta.summary	11/8/2014 5:35 PM	SUMMARY File	2 KB

圖 S4-8　　Counter/output_files 一覽

12-4　Synthesis FPGA 的介紹

FPGA Synthesis 首先要做的是 **File > Open Project**，如圖 S5-1 所示。

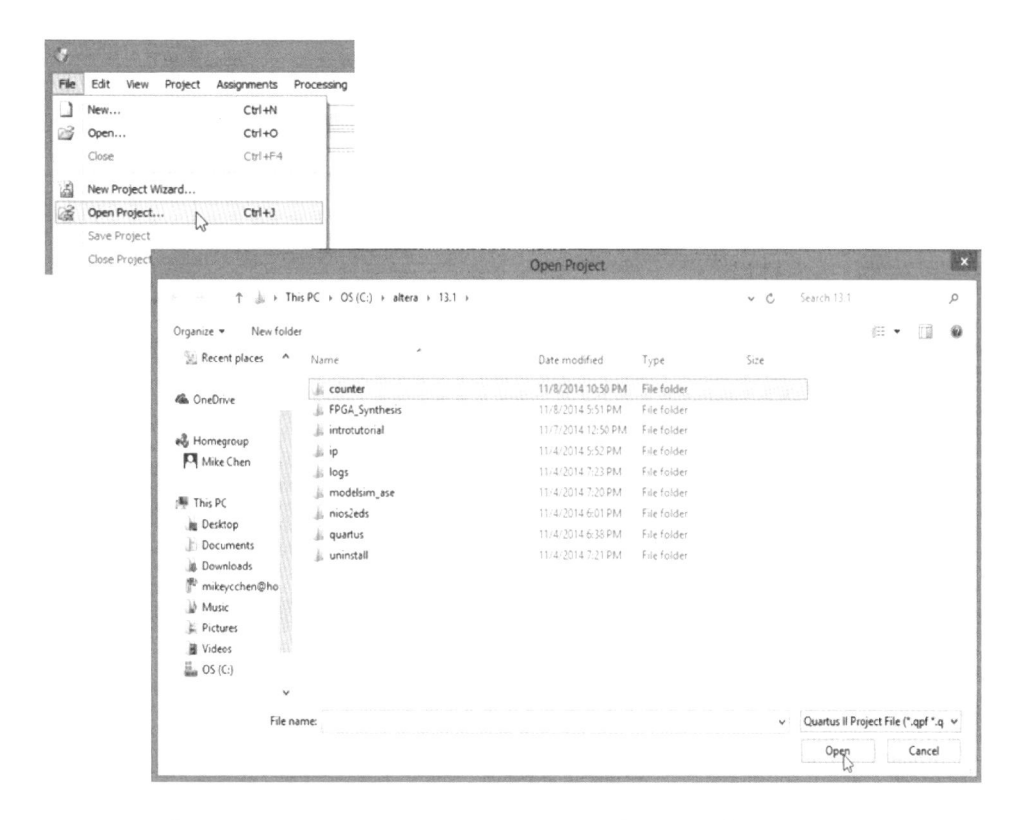

圖 S5-1　FPGA Synthesis 的第一步，啟動 Project

然後在 counter 的 File 中選擊 counter.qpf，如圖 S5-2 所示。

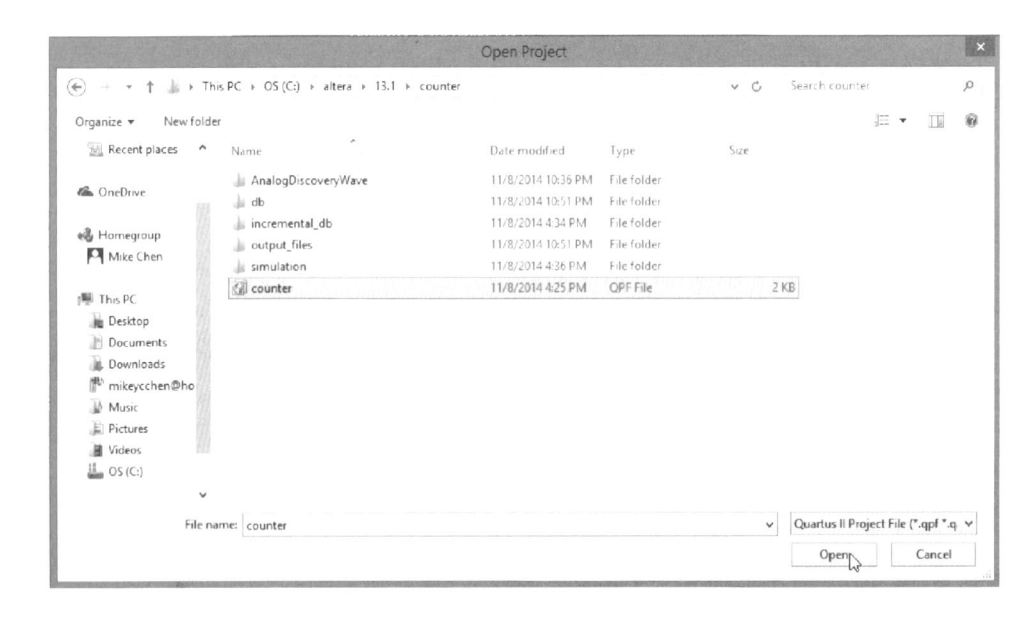

圖 S5-2　第二步開啓 counter.qpf

第十二章　Altera/Quartus DE2-115 FPGA 的電路合成

Project counter 便全部進入 Quartus II 軟體中，如圖 S5-3 所示。

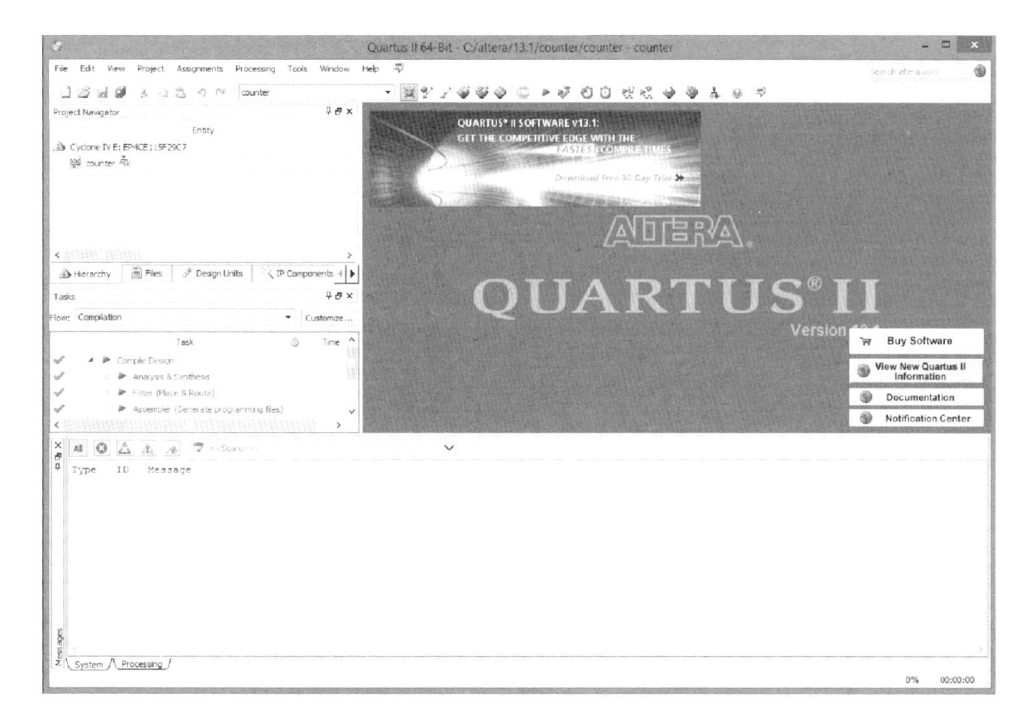

圖 S5-3　Project counter 進入 Quartus II

在 Programmer 進入工作之前，必須將 DE2-115 Power Up。而且它的 USB-Blaster 已經 Setup。然後再選用 **Tool > Programmer**，如圖 S5-4 所示。

然後選用 **Tools > Programmer**，如圖 S5-4 所示。

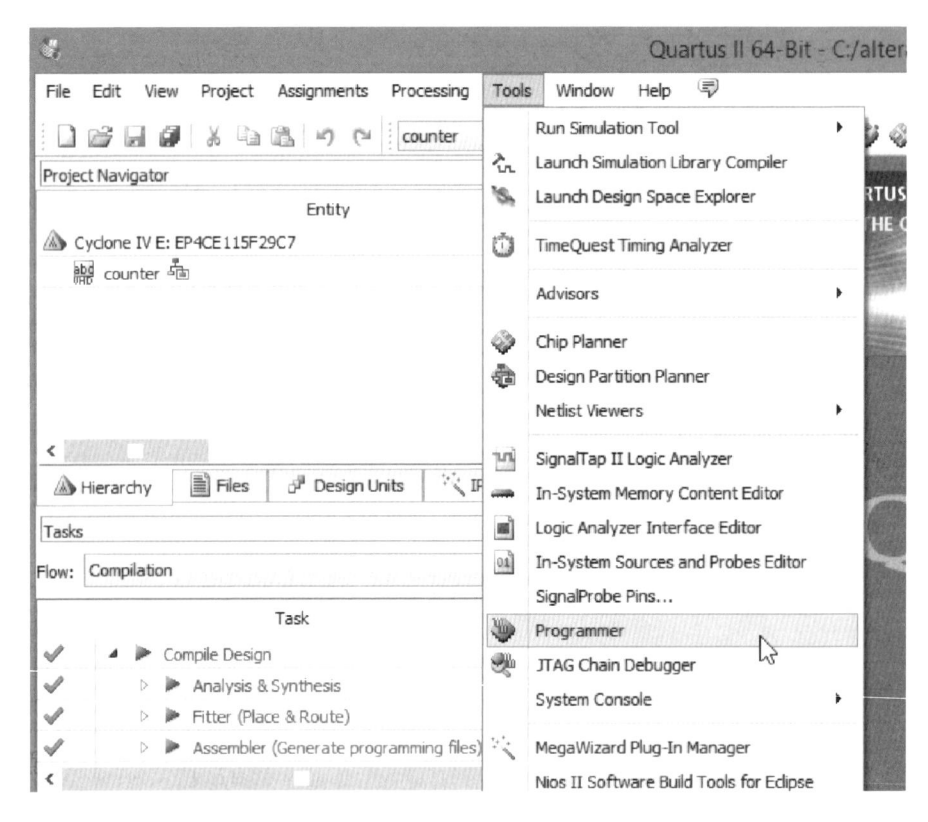

◎ 圖 S5-4 ◎　　　第四步選用 Tools > Programmer

第十二章　Altera/Quartus DE2-115 FPGA 的電路合成　155

圖 S5-5 顯示 USB-Blaster (USB-0)，已經設定。FPGA 的 Mode: JTAG，注意 DE2-115 的 RUN/Program 開關必須在 **RUN** 的位置。

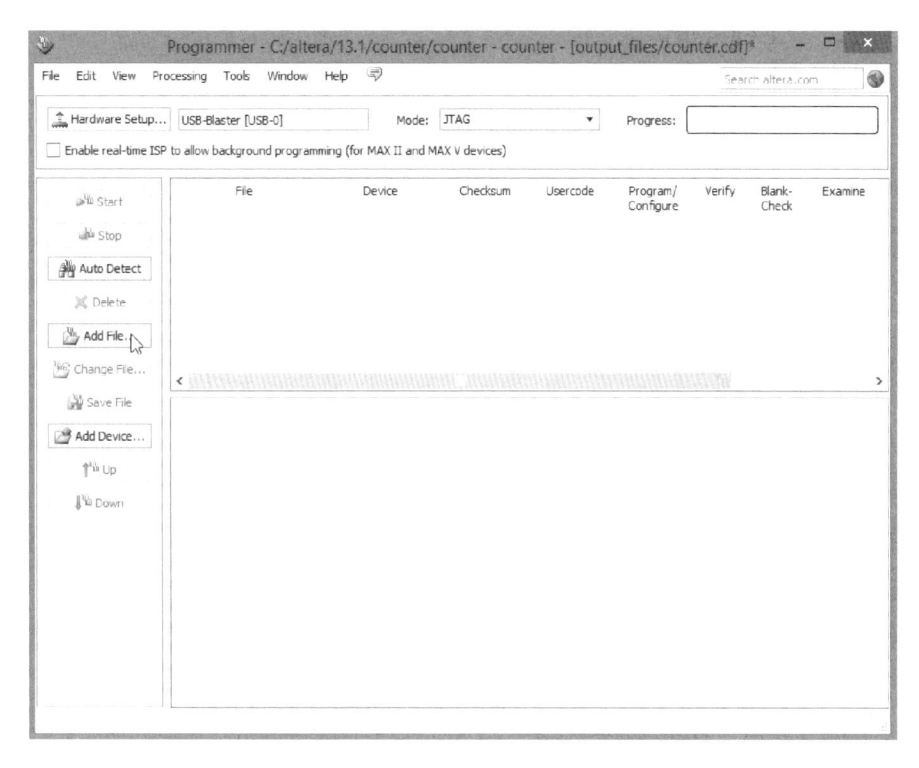

圖 S5-5　第五步檢查硬體設定

當一切合乎條件，單擊 **Add File**。當圖 S5-6 Select Programming File 視窗出現時，選用 **counter.sof File**。

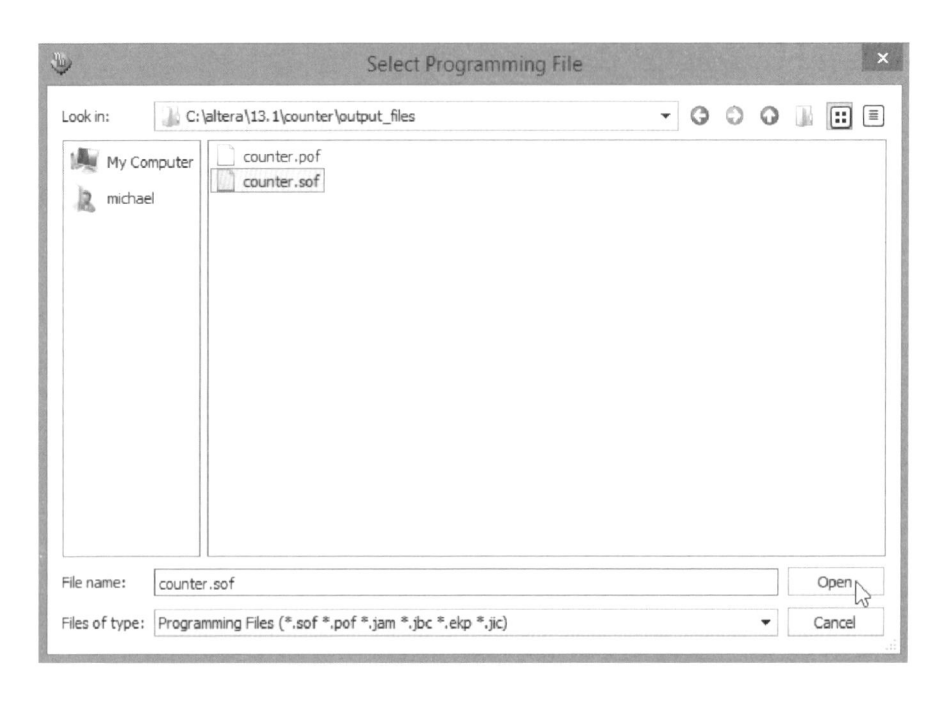

圖 S5-6　第六步選用 counter.sof File

第十二章　Altera/Quartus DE2-115 FPGA 的電路合成　157

這時候 FPGA/EP4CE115F29 和 counter.sof 自動顯示在圖 S5-7 中。

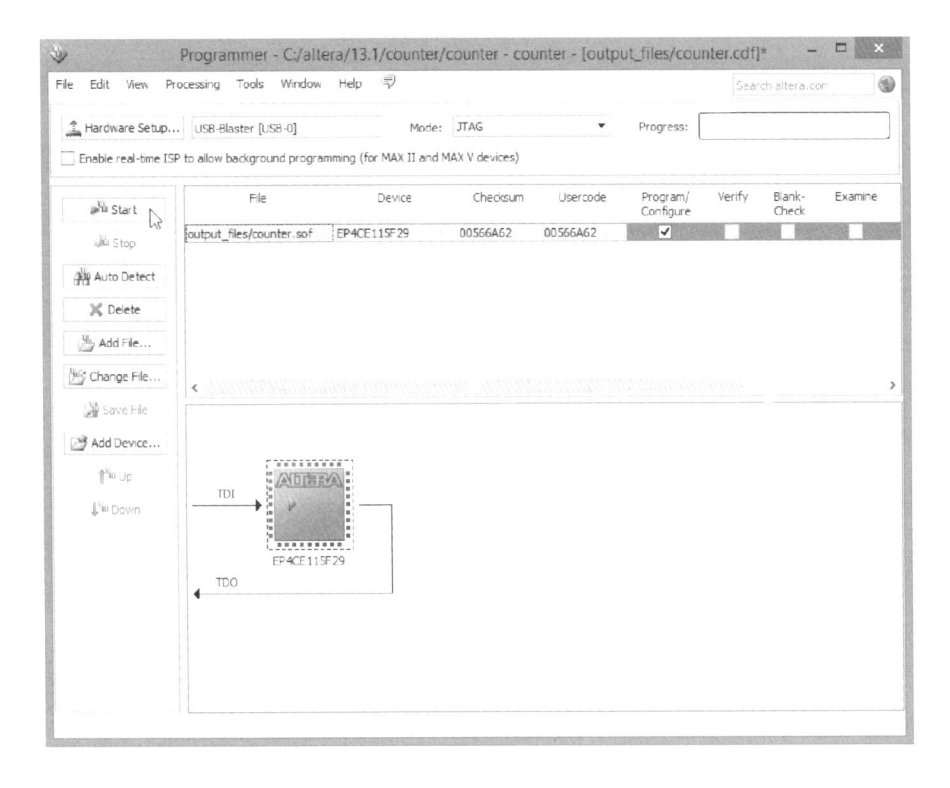

圖 S5-7　第七步開始 Programming FPGA

點擊圖 S5-7 左上角的 **Start**，如果一切正常 Progress 項當顯示 100% (**Successful**)，如圖 S5-8 所示，否則會用紅色來顯示 **Failed**。

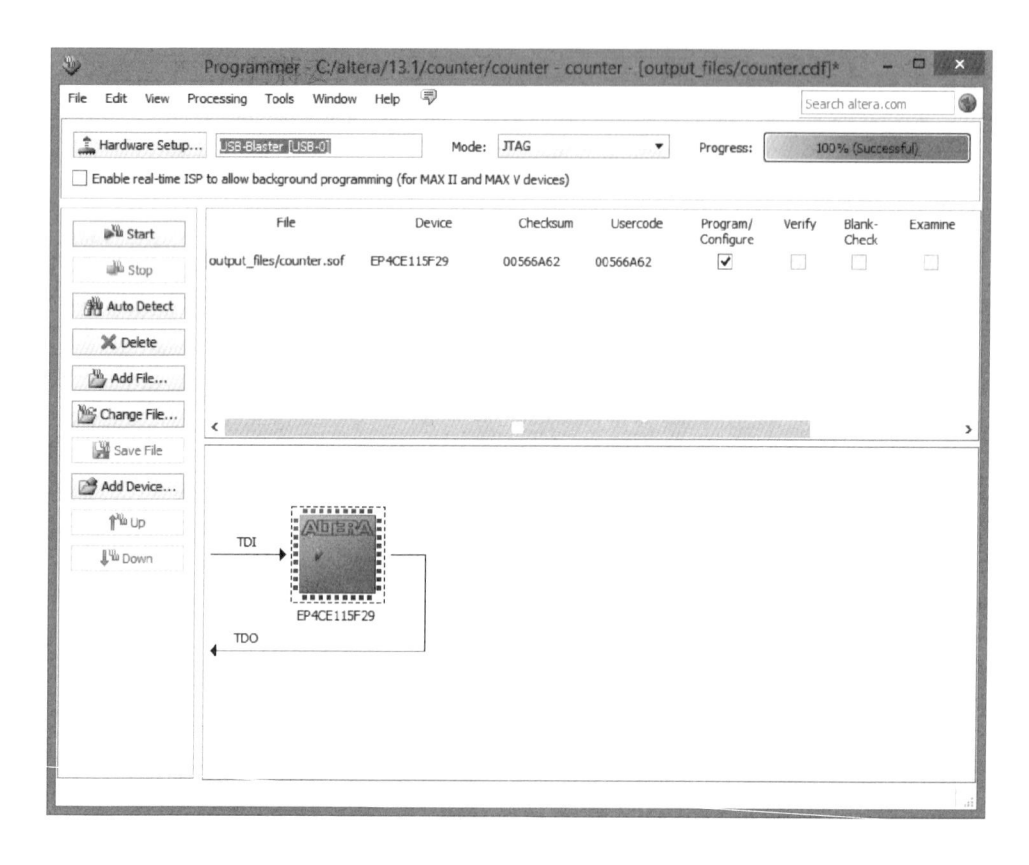

圖 S5-8　Programming FPGA 成功

第十二章　Altera/Quartus DE2-115 FPGA 的電路合成

Programming EEPROM/EPCS64，跟 FPGA 類似，所不同點為 Mode 要改成 Active Serial Programming。File 改成 counter.pof，Device 改為 EPCS64。同時選用 Verify、Blank Check 和 Program/Configure。如圖 S5-9 所示。

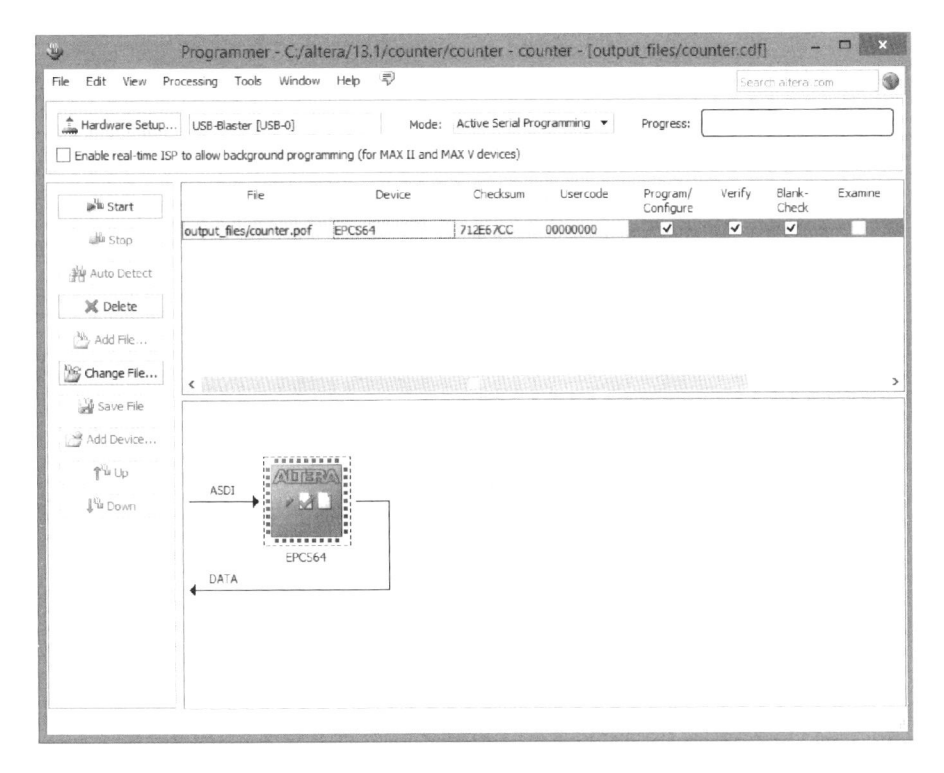

圖 S5-9　EEPROM/EPCS64 的 Programming 設定

DE2-115 的 RUN/PROGRAM 開關要放在 Program 上。當一切準備好了，單擊左上方的 **Start**，如果一切正常，Progress 項當顯示 100% **(Successful)**，如圖 S5-10 所示，否則會用紅色來顯示 Failed。

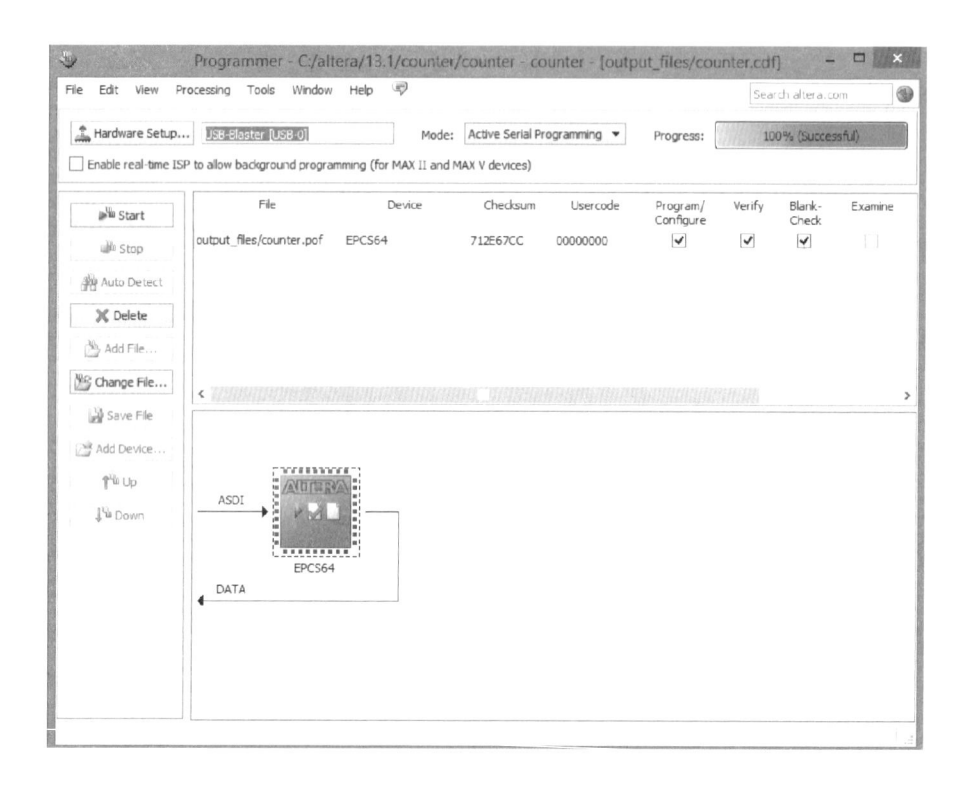

圖 S5-10　EEPROM/EPCS64 的 Programming 成功

12-5　Analog Discovery 對 DE2-115 FPGA 的測試

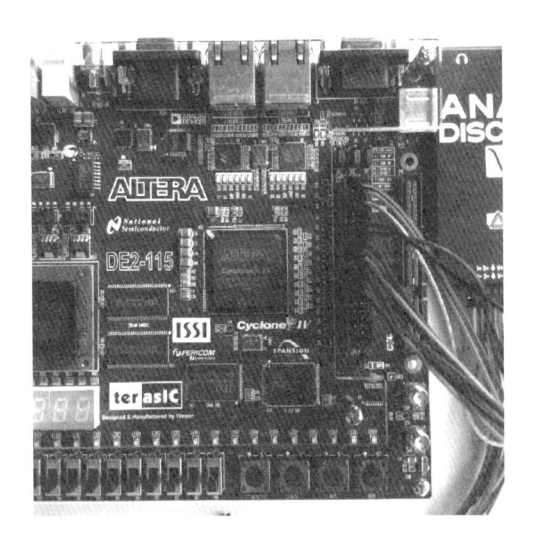

圖 S6-1　　DE2-115 JP5 與 Analog Discovery 的測試連線

　　Analog Discovery 的 Digital out 是一個 Digital Pattern Generator。Digital in 是一個 Logic Analyzer。這二部儀器，合用 16 條 I/O 接線。

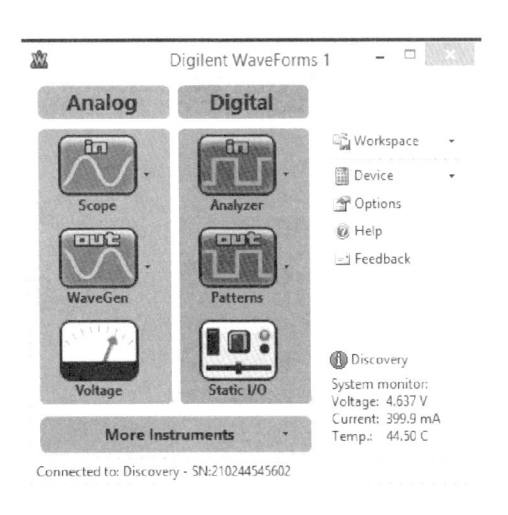

圖 S6-2　　開啟 Digital Waveforms 軟體

依據圖 S3-4，counter 的訊號。開啓並設定 Digital Pattern Generator，如圖 S6-3 所示。DIO 10 為 **clock**，DIO 9 為 **reset**，DIO 8 為 **load**，DIO 7 ~ DIO4 構成 Bus 0 = b000。Digital Pattern Generator 的 Trigger 設定為 Analyzer。

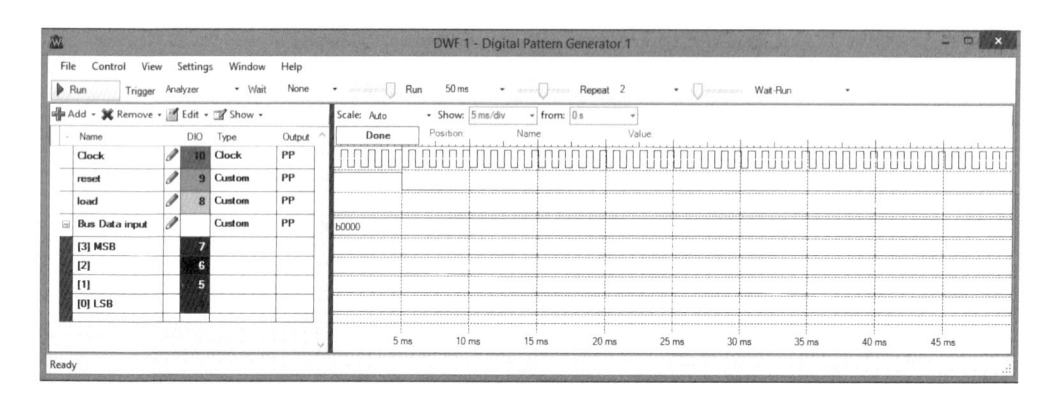

圖 S6-3　Digital Pattern Generator 對 counter.vhd 電路的設定

圖 S6-3 的 clock 頻率設定為 1 KHz，Trigger 為 Analyzer。Run time 為 50 ms。

圖 S6-4 的 Logic Analyzer，DIO 3~DIO 0 為 4 bit counter 的 4 個輸出。Mode 設定為 **Auto**。Source 設定為 **Analyzer**。

測試時，首先 Run Digital Pattern Generator，此時 Generator 波形的左上方會顯示 Armed。待點擊 **Analyzer** 的 **Single** 時，**Generator** 才會對 Analyzer 輸出預設的 **clock** 和 **reset** 訊號。同時 Analyzer 同步記錄到 4-bit counter 的 4 個輸出 DIO 3~DIO 0 的訊號波形。

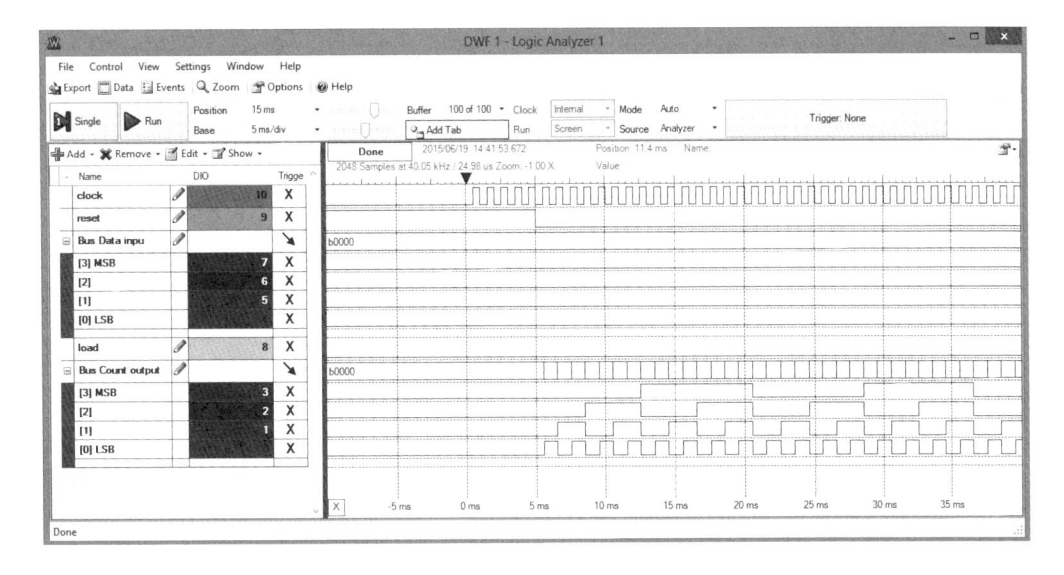

第十二章　Altera/Quartus DE2-115 FPGA 的電路合成

圖 S6-4： Logic Analyzer 記錄 4-bit counter 的 4 個輸出 DIO 3~DIO 0

12-6 課外練習

1. 試用 DE2-115 的 FPGA 來合成第二章的 Simple Adder，並用 Analog Discovery 的 Digital Pattern Generator 和 Logic Analyzer 來測試之。

2. 試用 DE2-115 的 FPGA 來合成第三章的 JKFF，並用 Analog Discovery 的 Digital Pattern Generator 和 Logic Analyzer 來測試之。

3. 試用 DE2-115 的 FPGA 來合成第四章的 3 to 8 Decoder，並用 Analog Discovery 的 Digital Pattern Generator 和 Logic Analyzer 來測試之。

4. 試用 DE2-115 的 FPGA 來合成第五章的 Shift Register，並用 Analog Discovery 的 Digital Pattern Generator 和 Logic Analyzer 來測試之。

第十三章　Xilinx/Vivado Basys3 FPGA 的電路合成

附錄 G 介紹了 Xilinx CPLD 的電路合成，使用的軟體為 ISE。ISE 的最後版本為 2013 年 10 月的 ISE 14.7。取而代之的為 Vivado，目下 Vivado 支援的 FPGA 為 Virtex-7、Kintex-7、Artix-7 和 Zynq-7000。Digilent 相應的 FPGA 教育用板子有 Arty、Basys3 和 Nexys4 等三種。這一章要介紹的是以第六章的 VHDL 計數器用 Vivado 軟體來合成到 Basys3 的 FPGA 板子上。

13-1　Vivado 軟體的運作介紹

　　Vivado 軟體與 ISE 軟體不同，它把 Project 所需的各種工具包羅在內。HDL simulator 便是其中的一個例子。由於作業的一貫性，因此可以節省操作的時間，並且隨時可以從項目的 Summary 裡查看到由各個工具運作後所產生的資料。

　　以第六章的 Counter 為例，用 Vivado 對 Basys3/FPGA 的合成，除了需要 Counter 的 Device file "counter.vhd"，和 Simulation 所需的 "testbench.vhd"，另外再加上 counter 在 Basys3 上的接腳檔 "pin.xdc"。完整的 Basys3 接腳檔由 Digilent 所提供，名為 Basys3_Master.xdc。

啓動 Vivado 後 Windows 8.1 所顯示的畫面如圖 13-1

◦ 圖 13-1 ◦　　　啓動 Vivado 後 Windows 8.1 所顯示的畫面

使用 Basys3/FPGA 來合成 counter，點擊圖 13-1 的 Create New Project 便有圖 13-2 的 Create a New Vivado Project > Next。然後在選定 Project location 後把 Project Name 上填入 counter。

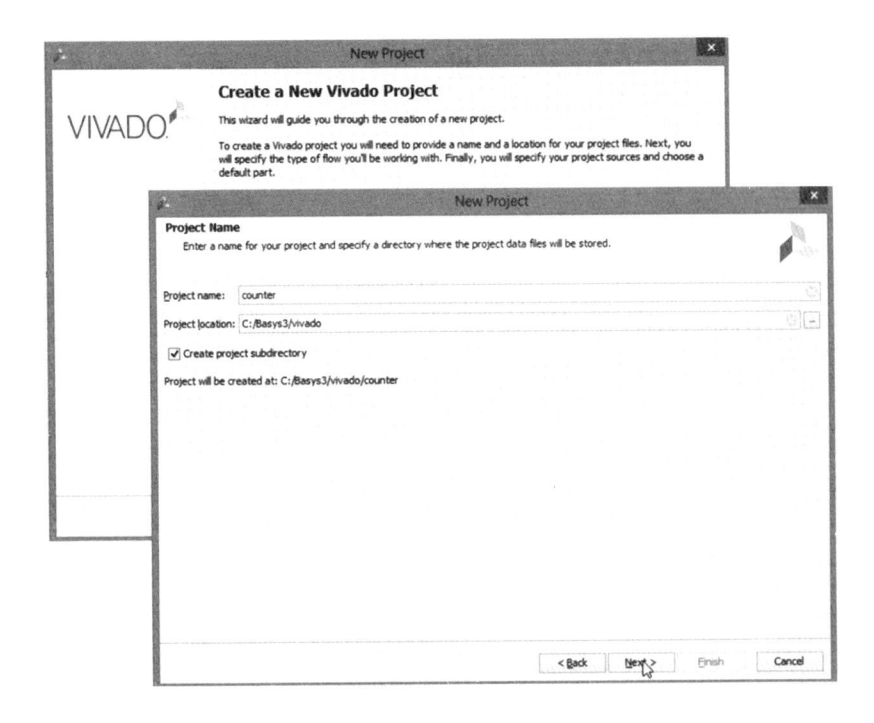

第十三章　Xilinx/Vivado Basys3 FPGA 的電路合成　167

◦ 圖 13-2 ◦　　選定 Project location 後在 Project Name 上填入 counter

圖 13-2 為 Project Type，選用 RTL > Next。然後選用 Basys3 上的 FPGA。

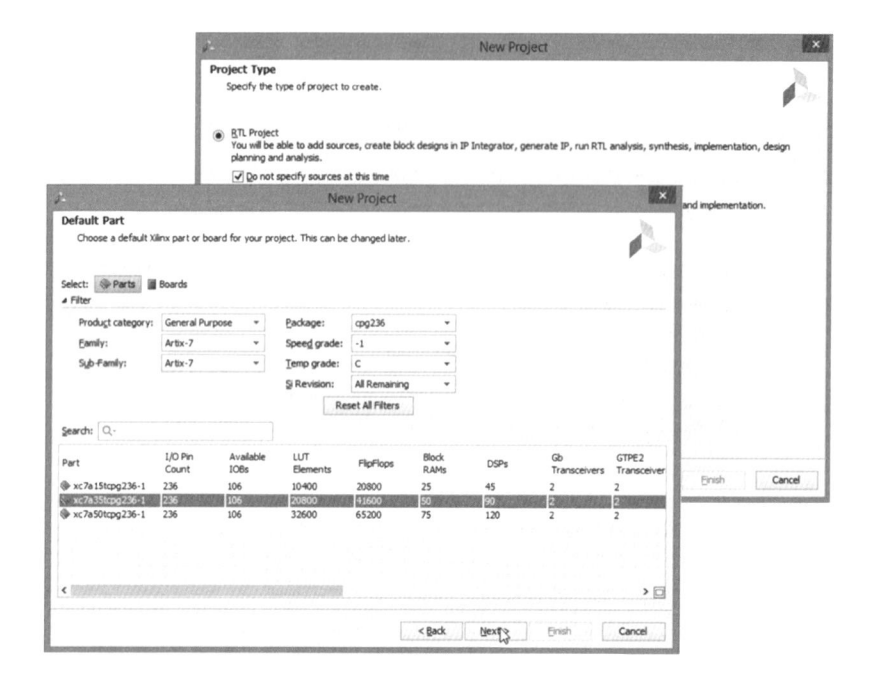

圖 13-3　選用 Project 的 Type 及 Basys3 上的 FPGA

第十三章　Xilinx/Vivado Basys3 FPGA 的電路合成　169

圖 13-4 為以上所完成的概要 New Project Summary。

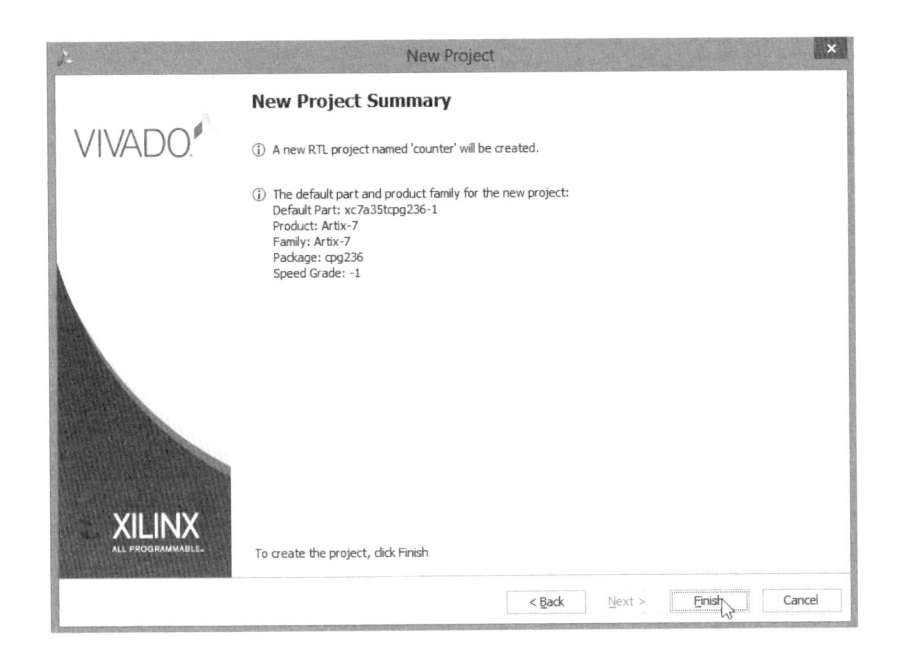

圖 13-4　New Project 的結論

接下來要做的是將 counter.vhd、pin.xdc 和 testbench.vhd 相繼加入到圖 13-5 中

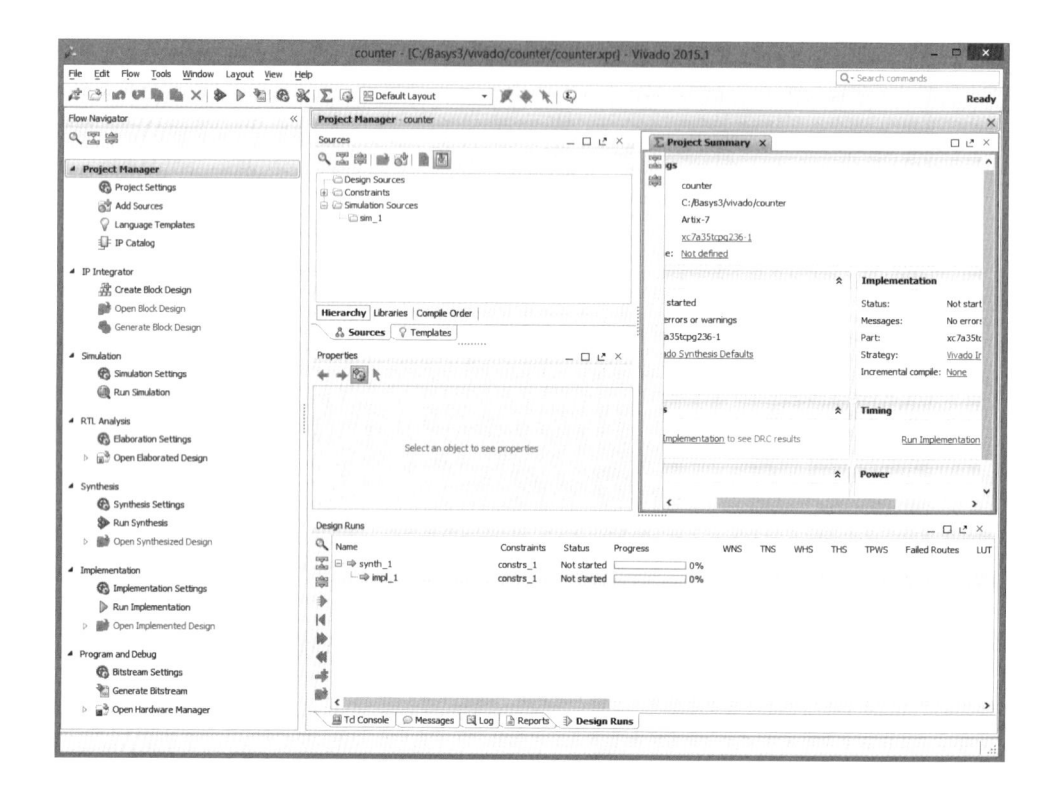

圖 13-5：Vivado 的 Flow Navigator 視窗

圖 13-5 的左邊 Flow Navigator 內包羅了 Project Manger、IP Integrator、Simulation、RTL Analysis、Synthesis、Implementation、和 Program and Debug。

第十三章　Xilinx/Vivado Basys3 FPGA 的電路合成

要把 counter.vhd 加入到 New Project 如圖 13-6 在 Flow Navigator 的 Project Manger 中單擊 Add Source。再在 Add Source windows 中選用 Add or create design source 然後 Next。

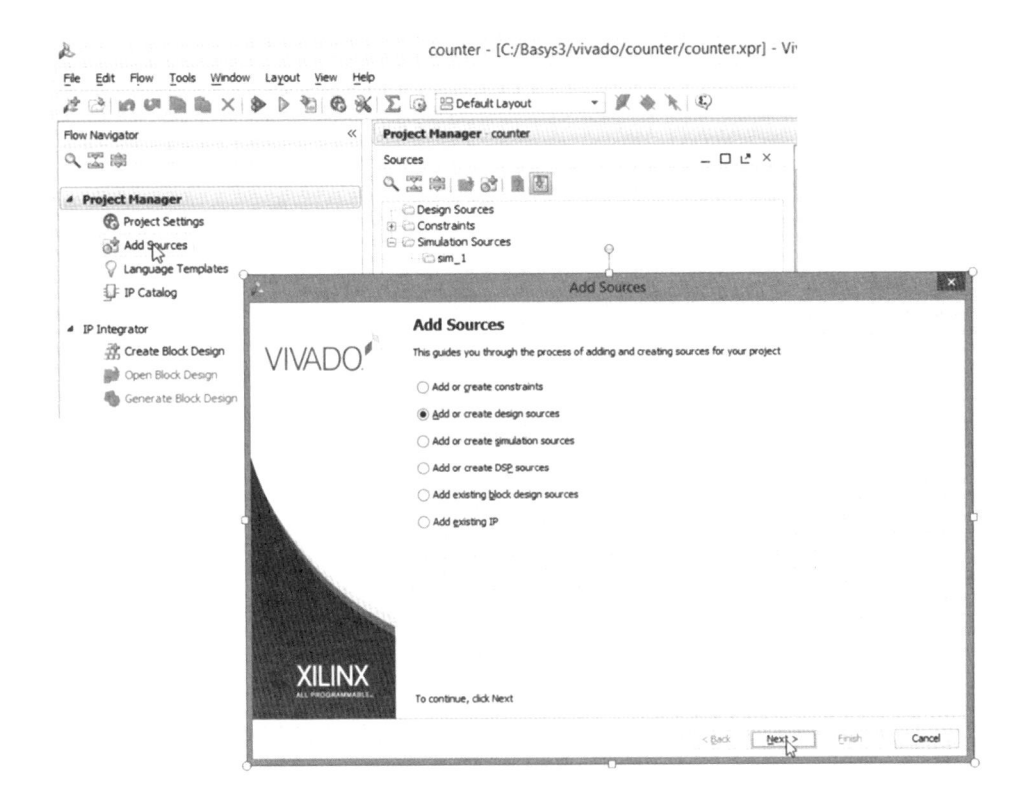

圖 13-6： 把 counter.vhd 加入到 New Project 步驟之一

接下來是加入第六章中已存在的 counter.vhd。如圖 13-7 所示。

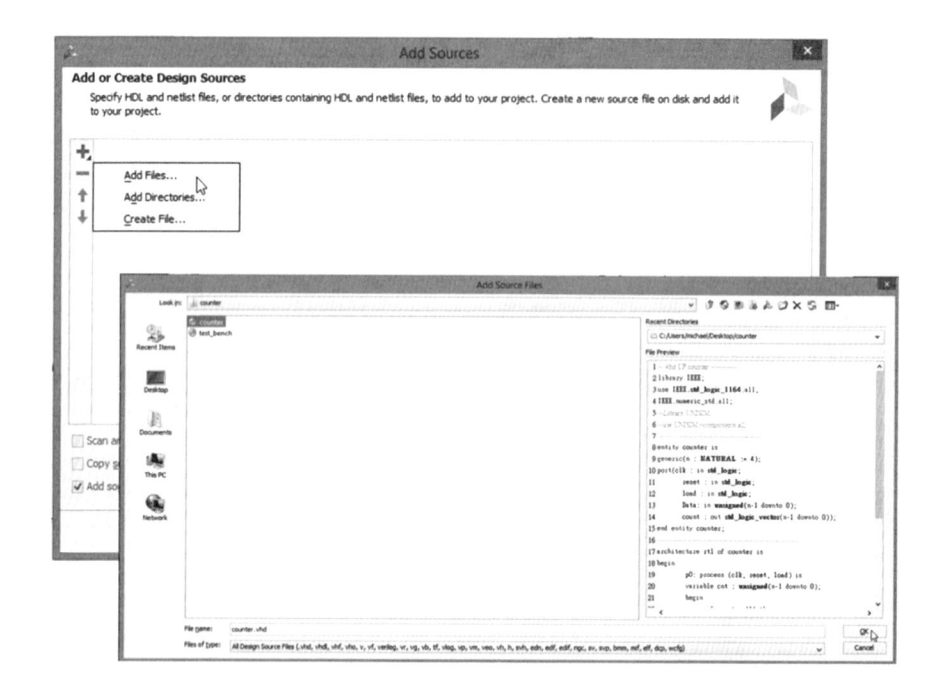

圖 13-7　把 counter.vhd 加入到 New Project 步驟之二

第十三章　Xilinx/Vivado Basys3 FPGA 的電路合成　173

圖 13-8 顯示 counter.vhd 已經加入，接下來再加入由 Basys3_Master.xdc 選項出來的 pin.xdc。

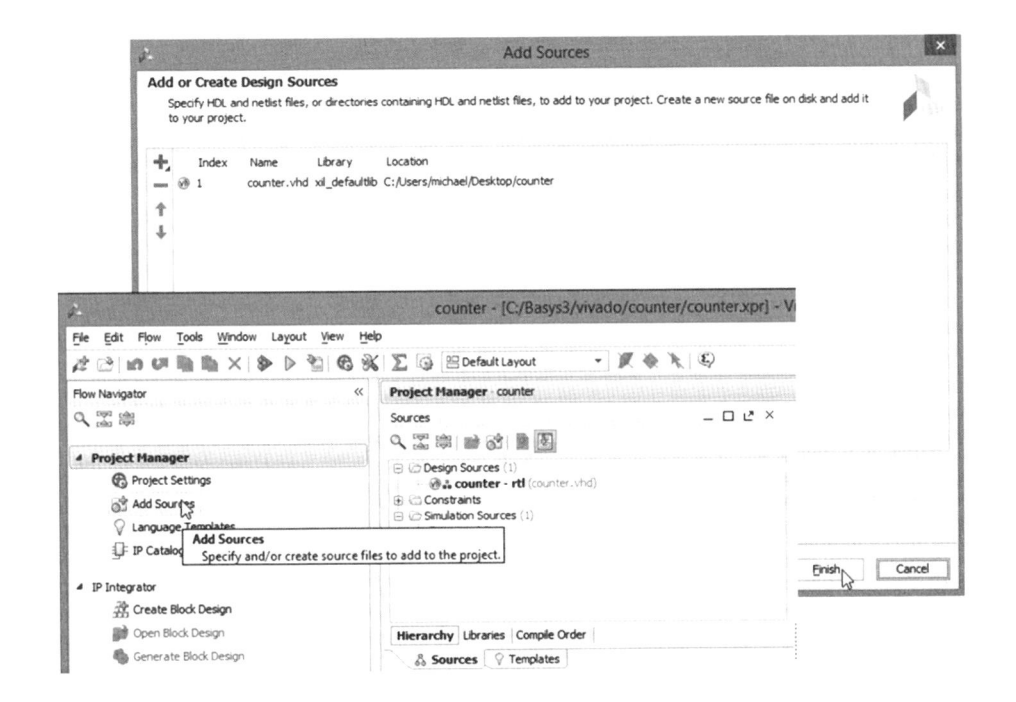

圖 13-8　把 counter.vhd 加入到 Project 步驟之三

同理如圖 13-9 所示，將 pin.xdc 加入到 Project 內。

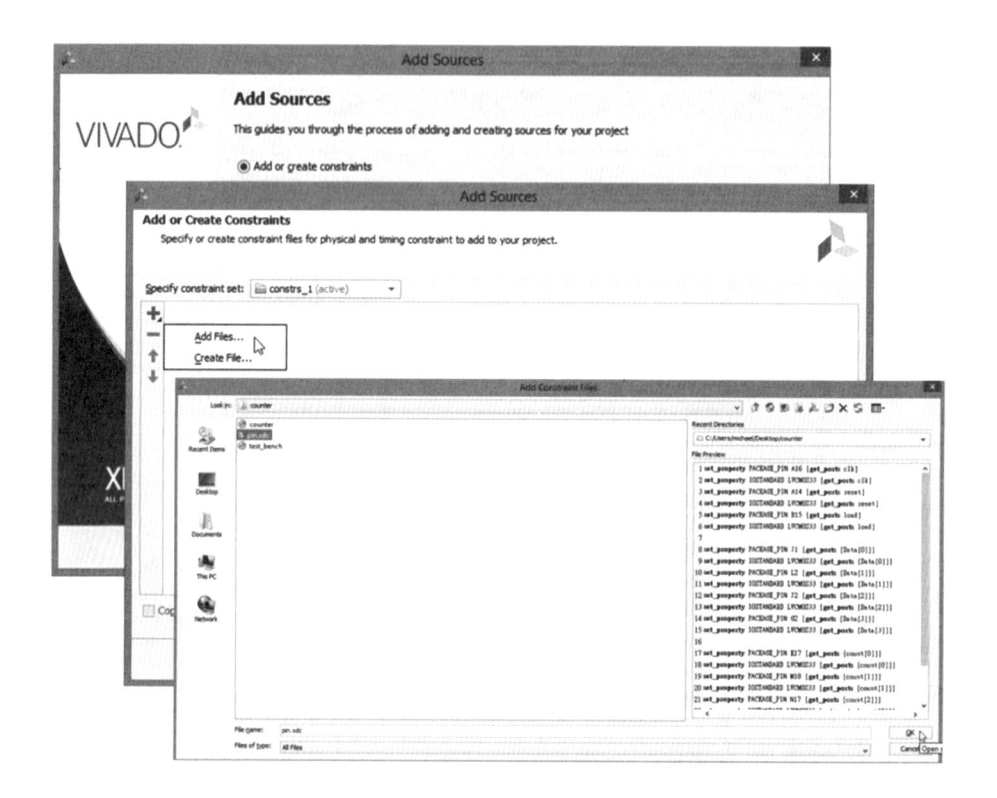

圖 13-9　把 pin.xdc 加入到 Project 步驟之一

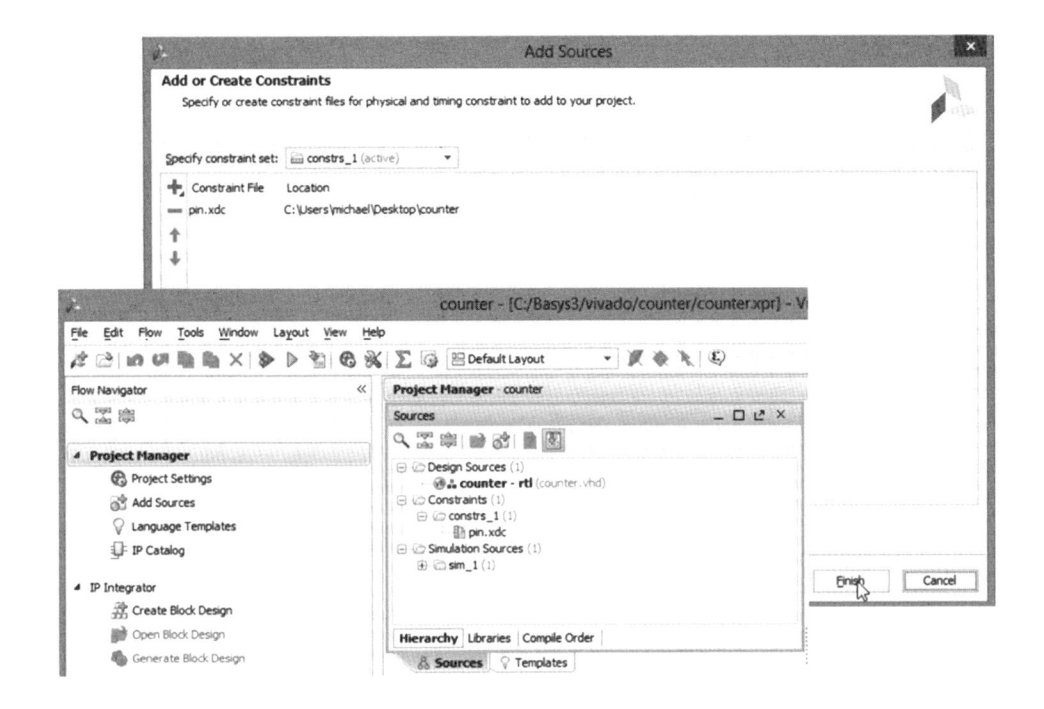

第十三章　Xilinx/Vivado Basys3 FPGA 的電路合成　175

▎圖 13-10 ▎　　　把 pin.xdc 加入到 Project 步驟之二

有了 counter.vhd 和 pin.xdc 檔之後便可在 RTL Analysis 項目下 Open Elaborate Design 如圖 13-11 所示。

◎ 圖 13-11 ◎　　在 RTL Analysis 項目下 Open Elaborate Design

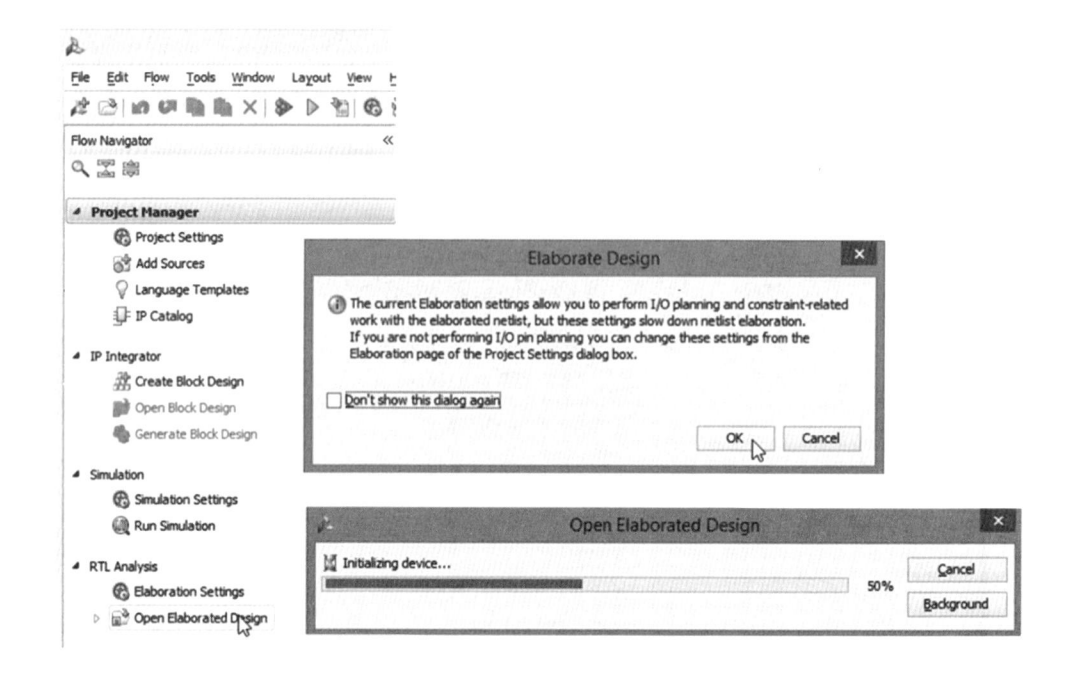

Open Elaborate Design 主要的是要觀察 counter 配上接腳的電路圖。如圖 13-12 所示。

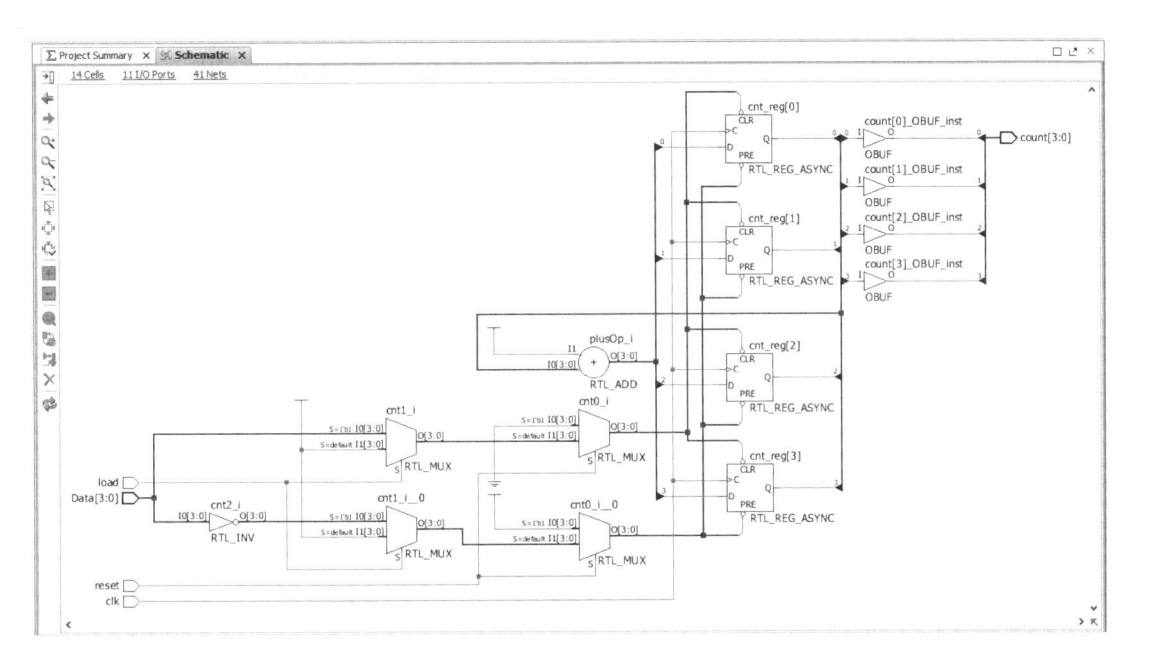

第十三章 Xilinx/Vivado Basys3 FPGA 的電路合成

圖 13-12　Open Elaborate Design 以觀察 counter 配上接腳的電路圖

13-2　Vivado 的 Simulator 運作介紹

　　為了要模擬測試 counter.vhd，New project 須要加入 testbench.vhd。如圖 13-13 同理在 Project Manager 中點擊 Add Sources，Add Sources 視窗中選　Add or create simulation sources　然後點擊 Next。

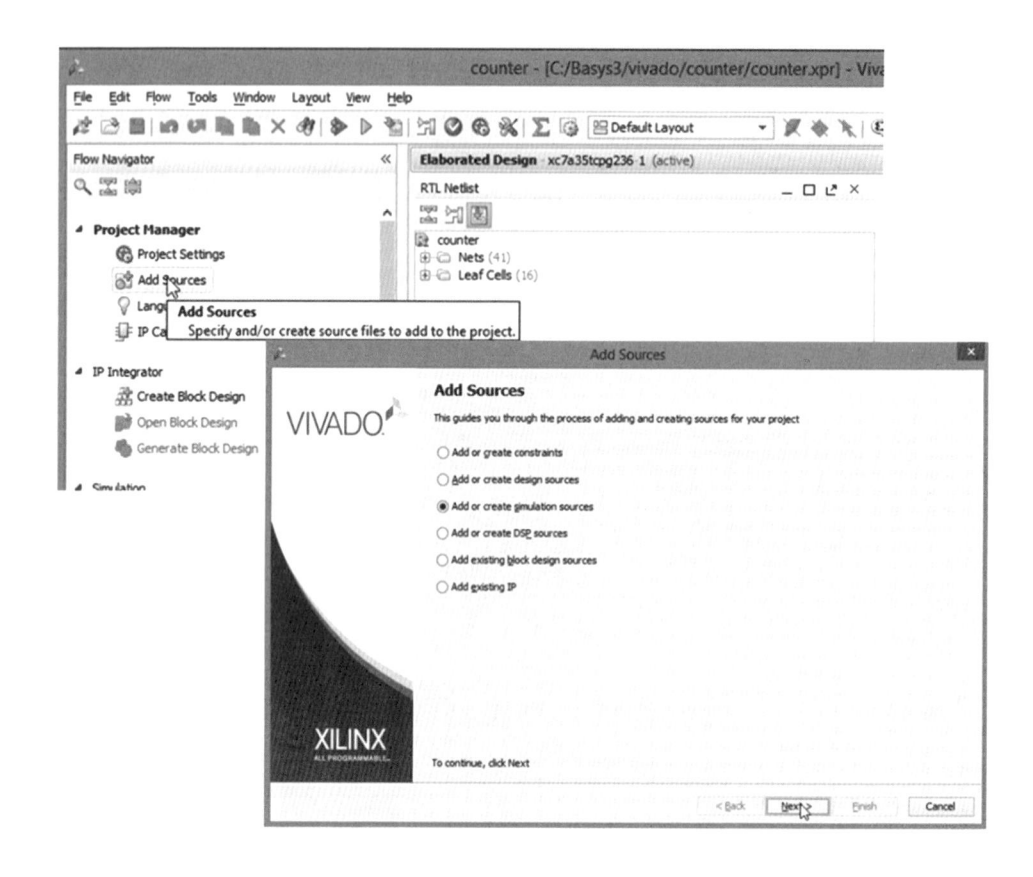

◎ 圖 13-13 ◎　　加入 simulation 所需的 testbench.vhd 步驟之一

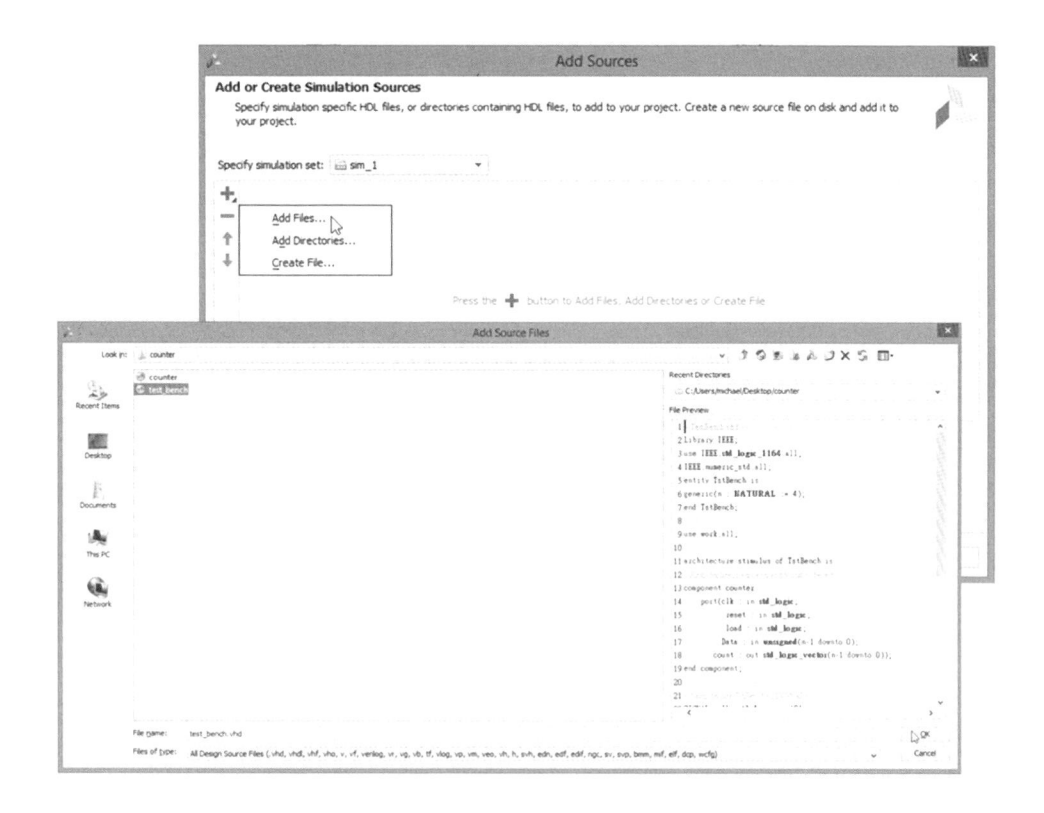

第十三章　Xilinx/Vivado Basys3 FPGA 的電路合成　179

圖 13-14：加入 simulation 所需的 testbench.vhd 步驟之二

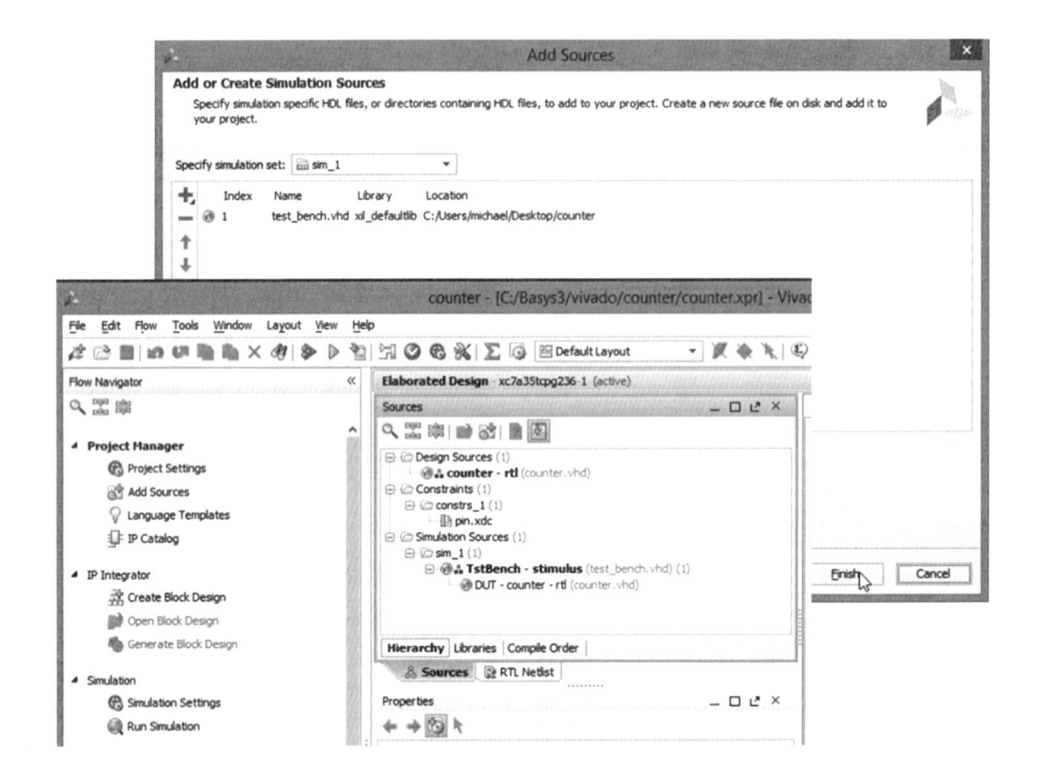

圖 13-15 加入 simulation 所需的 testbench.vhd 步驟之三

第十三章　Xilinx/Vivado Basys3 FPGA 的電路合成

　　Project 有了 Counter.vhd 和 Testbench.vhd 之後便可以著手 Simulation 的設定如圖 13-16 所示。在 Simulation 的選項中點擊 Simulation Settings。

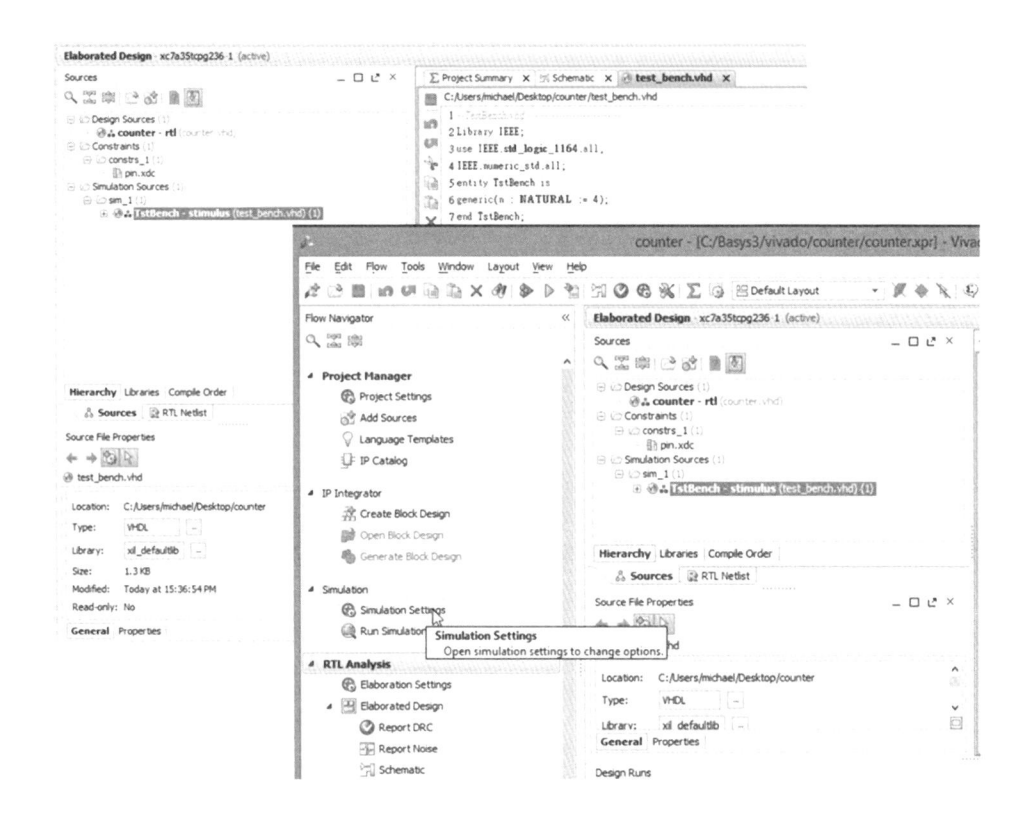

圖 13-16　　New Project 模擬測試的設定之一

如圖 13-17 所示填好 Vivador Simulator、VHDL、Testbench 點擊 OK。

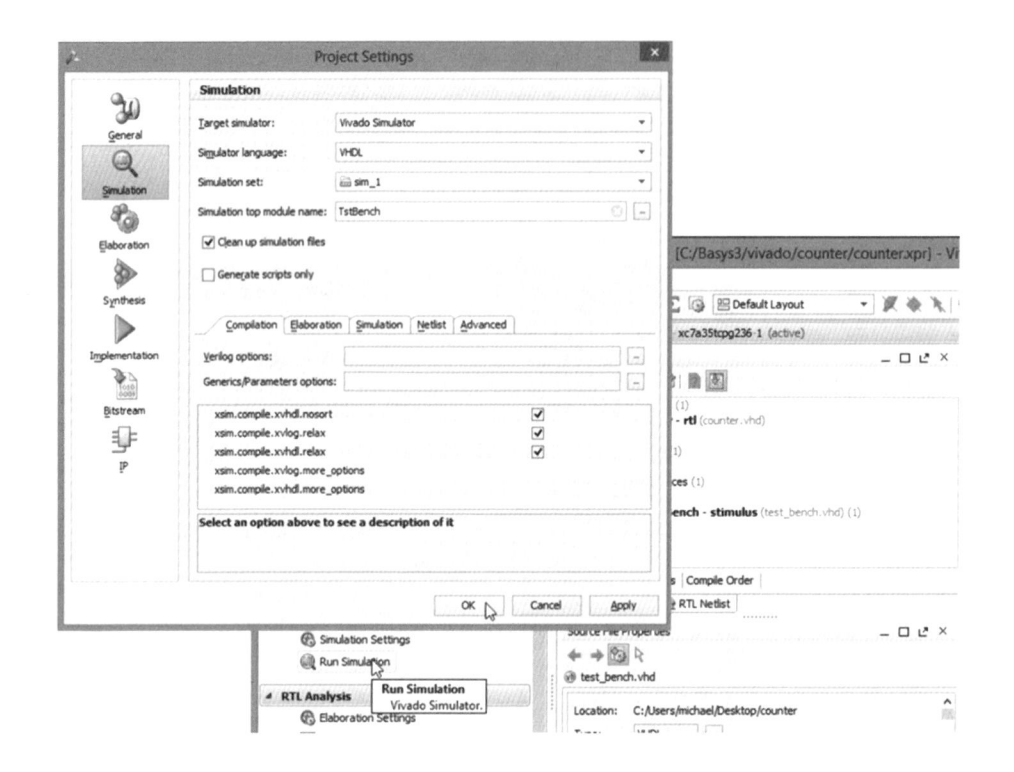

圖 13-17　　New Project 模擬測試的設定之二

第十三章 Xilinx/Vivado Basys3 FPGA 的電路合成

設定完畢之後，再點擊圖 13-18 中的 Run Simulation > Run Behavioral Simulation 結果如圖 13-19 所示。

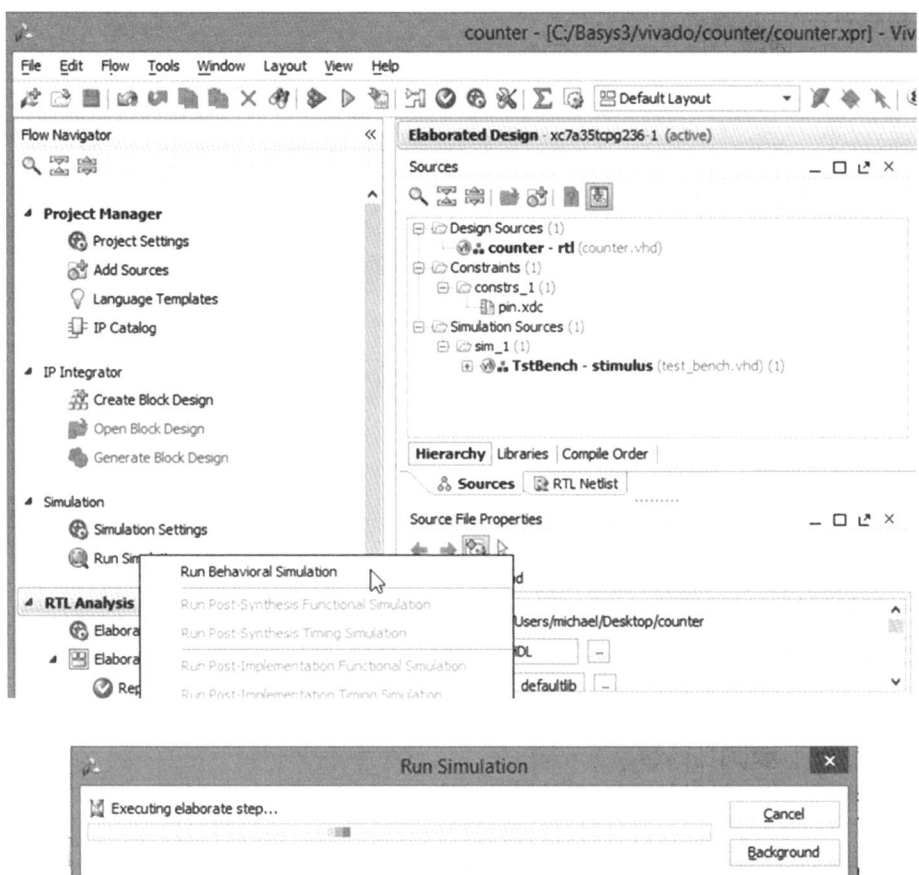

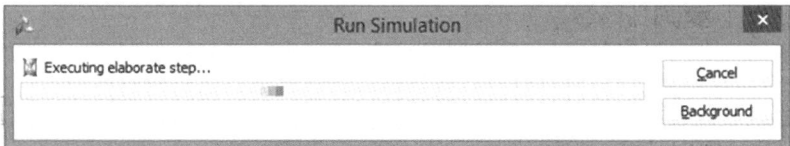

圖 13-18　Counter Project 進行 Run Behavioral Simulation

184　iLAB Digital 數位電路設計、模擬測試與硬體除錯

　　圖 13-19 為 Run Behavioral Simulation 的結果，用上、中、下三個畫面來敘述觀察波形的方法，最上面的畫面是 Run 多少個 ns。中間的畫面是將全部時間的畫面顯示出來。下面的畫面是把整個畫面展開來察看。

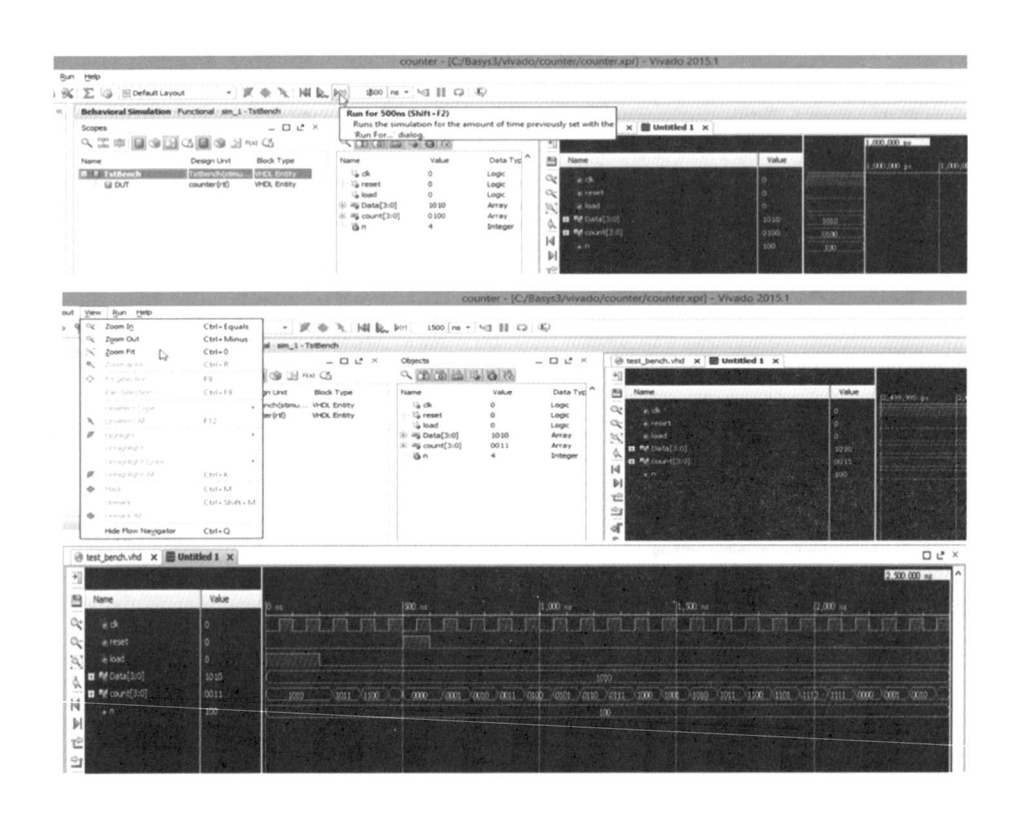

◦ 圖 13-19 ◦　　　Run Behavioral Simulation 的結果和觀察的方法

13-3　Vivado 軟體的 Implementation 和 Bitstream 檔案的產生

Simulation 成功之後就可以進行 Synthesis，如果沒有錯誤接下來可以 Run Implementation。如圖 13-20 所示。

圖 13-20　Synthesis 成功 Implementation 進行中

Implementation 成功之後便可以開啓來觀察，如圖 13-21 所示。

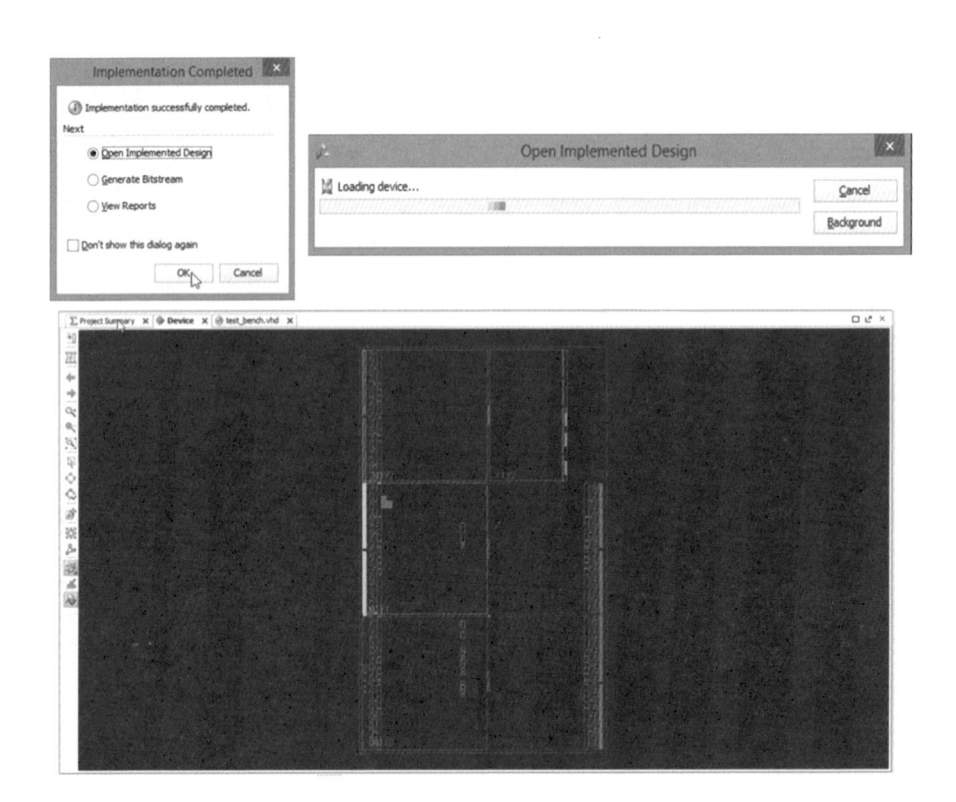

圖 13-21　Open Counter Project Implementation

同時可從 Project Summary 察看到整個 Project 開始到目前的工作記錄，如圖 13-22 所示。

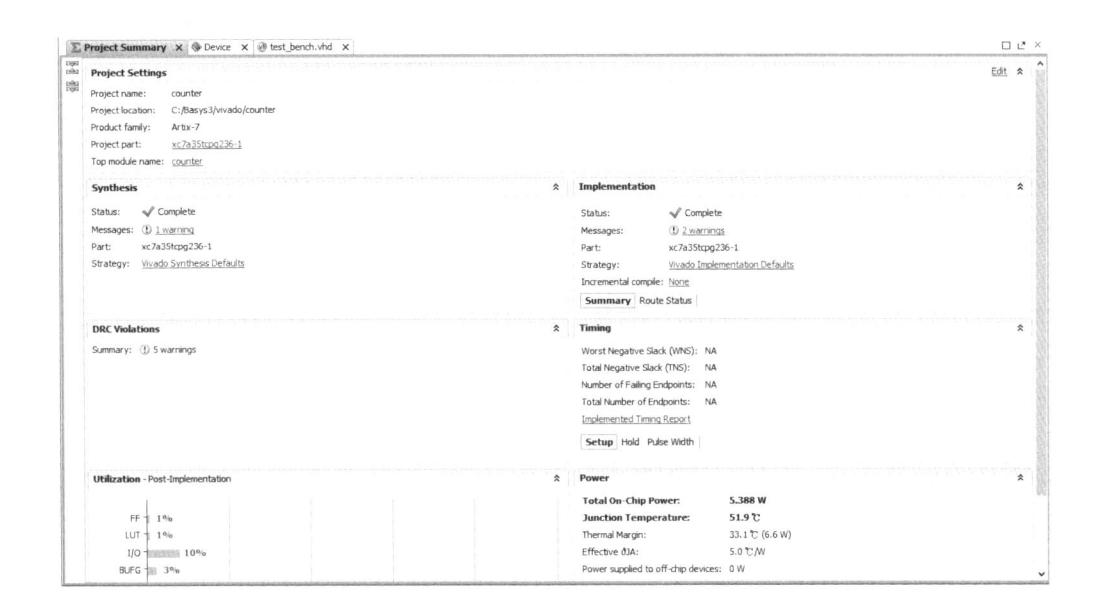

◎ 圖 13-22 ：　Project 開始到目前的工作記錄 Project Summary

188　iLAB Digital 數位電路設計、模擬測試與硬體除錯

　　點擊 Synthesis 中的 Open Synthesis Design，當 Synthesis windows 出現後如圖 13-23 所示，點擊其中的 Schematic 便可獲得如圖 13-24 的 FPGA/counter 的電路圖。

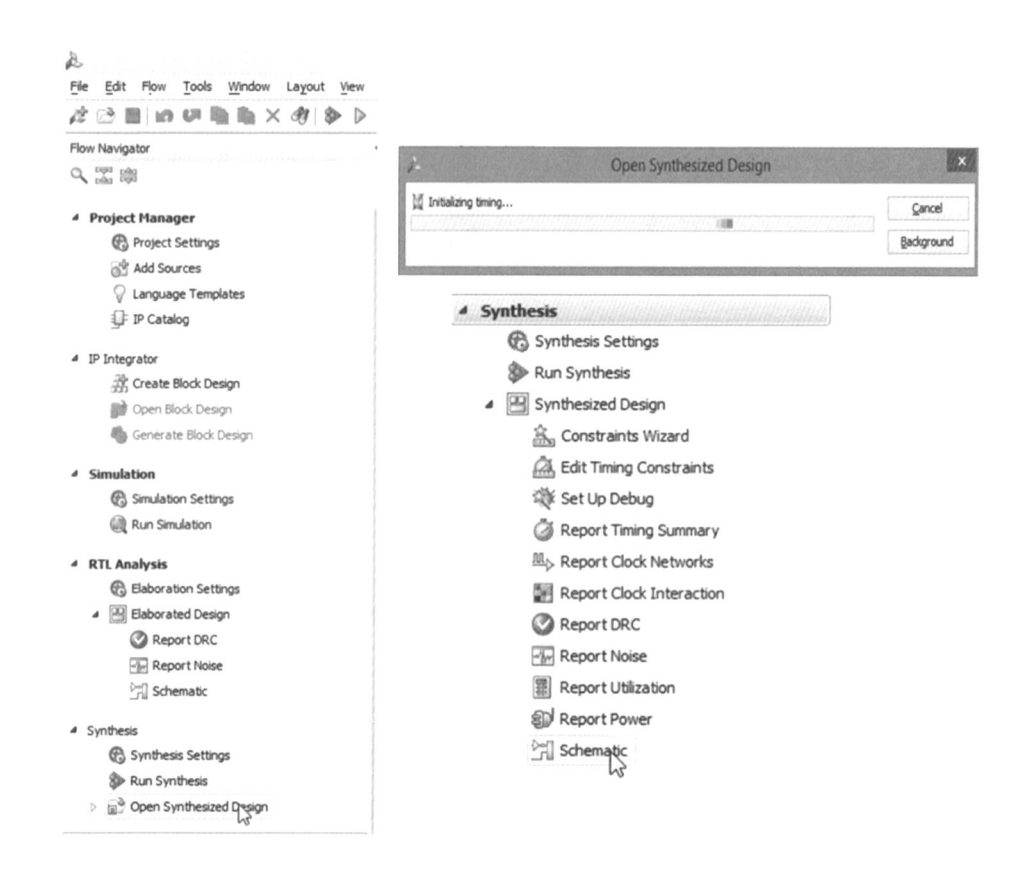

圖 13-23　點擊 Open Synthesis Design 和 Synthesis windows 出現

第十三章 Xilinx/Vivado Basys3 FPGA 的電路合成

點擊圖 13-23 中 Synthesis windows 的 Schematic 就會顯示 FPGA 所合成的 counter 電路圖，如圖 13-24 所示，它跟圖 13-12 所顯示的在結構上很不相同。

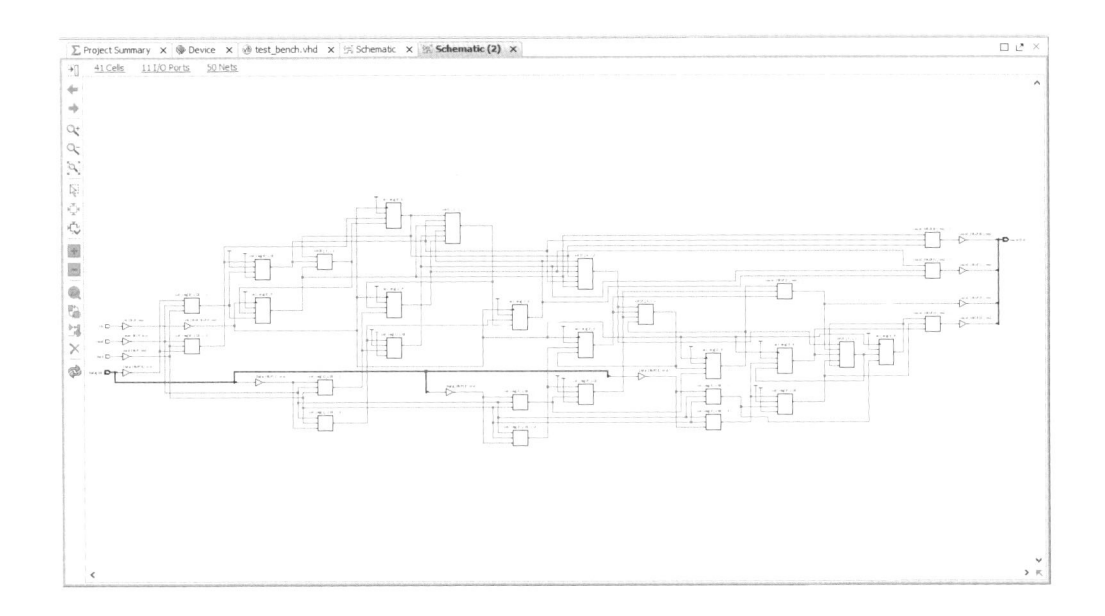

圖 13-24： Counter Project 所得之 FPGA /counter 的電路圖

Program FPGA 所需的 Bitstream 檔案，可從點擊 Program and Debug 中的 Generate Bitstream 入手，如圖 13-25 所示。

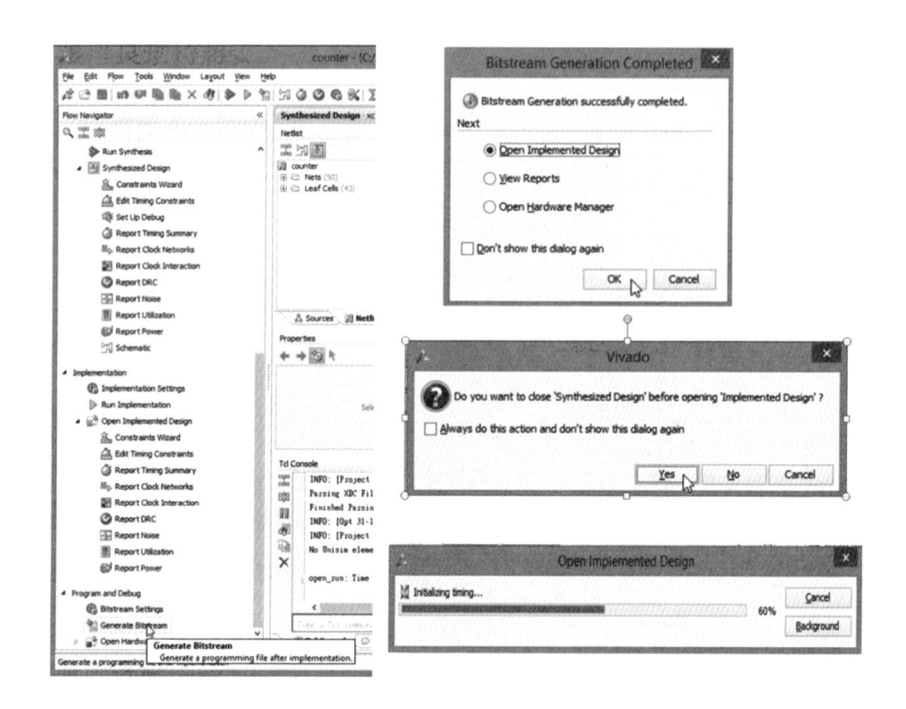

圖 13-25　　Bitstream 檔案的產生和 Open Implemented Design

13-4　Vivado 軟體的 Bitstream Programming Basys3

從 Program and Debug 的選項中 Open Hardware Manager，並且連接上 Basys3 到 PC 上，讓它們經 Driver 連接起來，如圖 13-26 所示。

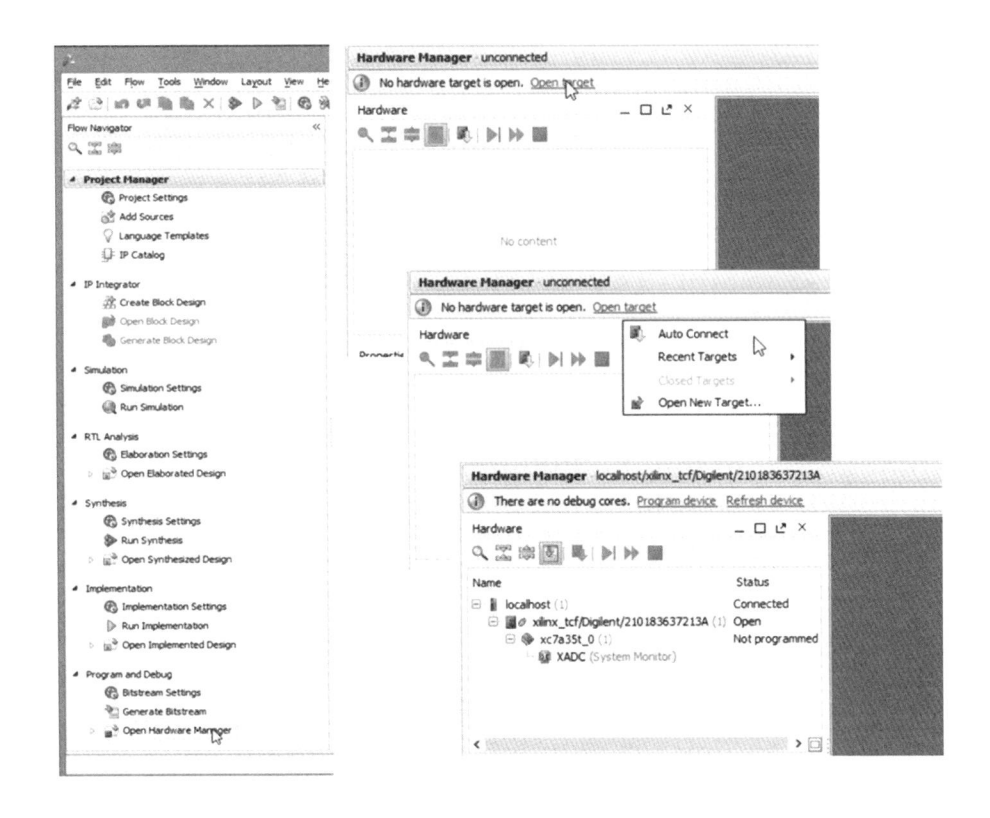

圖 13-26　Open Hardware Manager 使 Driver 連接 Basys3 到 PC 上

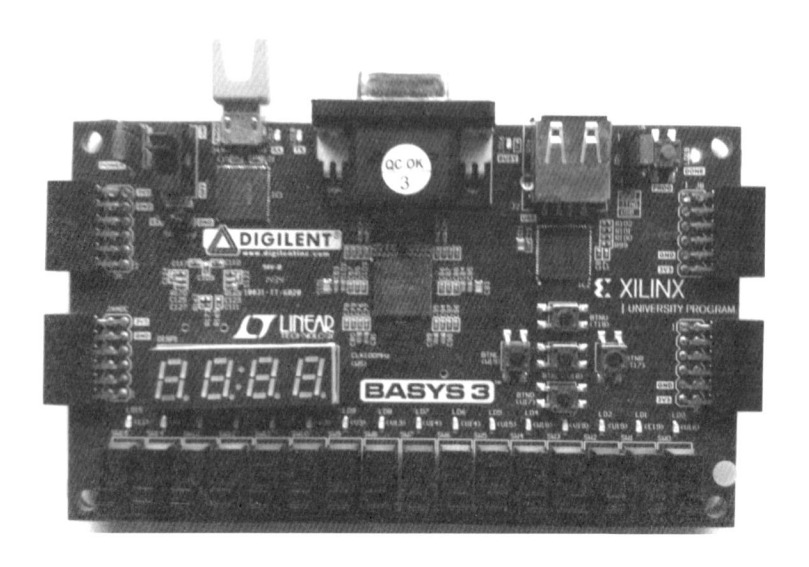

圖 13-27　Basys3 右上角的 Green LED 發亮表示 Programming 成功

第十三章　Xilinx/Vivado Basys3 FPGA 的電路合成　193

13-5　Analog Discovery 對 Basys3/FPGA Counter 的測試

依據 pin.xdc 檔的設定連接 Analog Discovery 的 Digital 測試線到 Basys3 的 3 個 Pmod connectors 上如圖 13-28 所示。

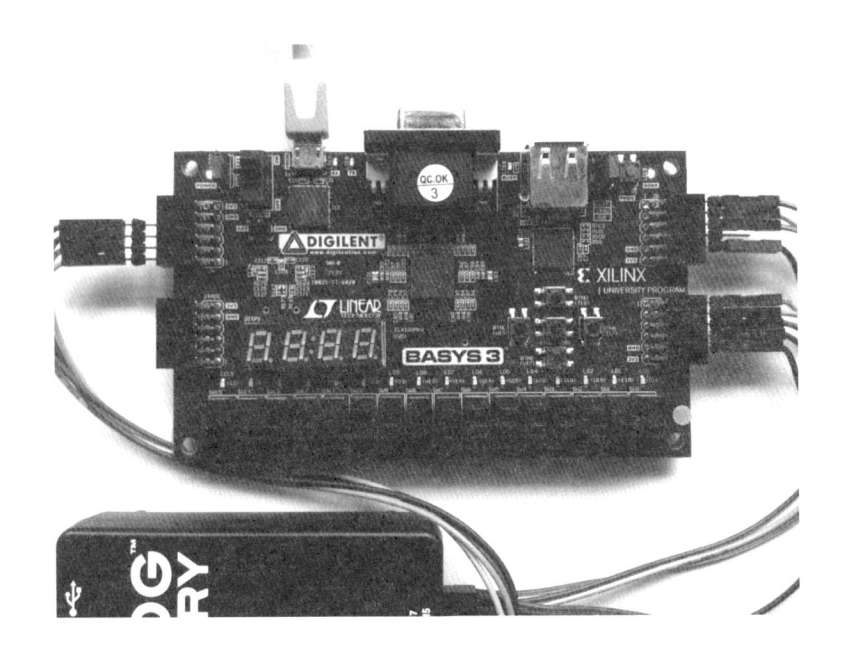

圖 13-28　依據 pin.xdc 檔的設定連接 Analog Discovery 到 Basys3 的 Pmods 上

在 Waveform1 的 Digital Pattern Generator 設定所有的輸入訊號，如圖 13-29 所示。

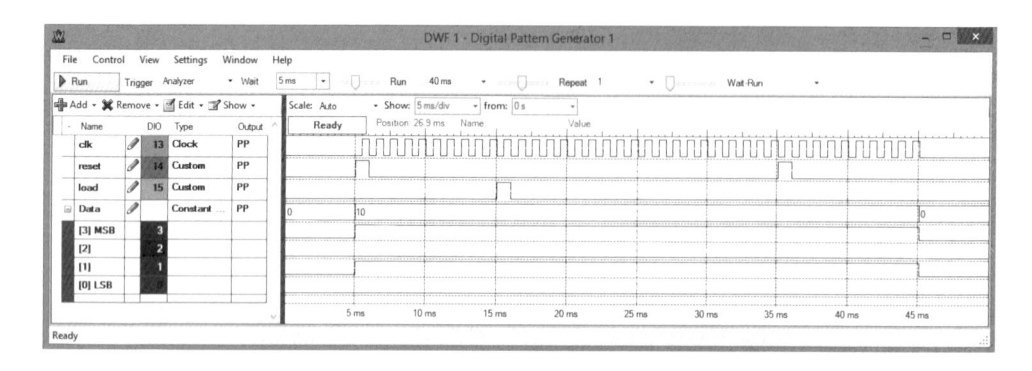

◌ 圖 13-29 ◌　　Digital Pattern Generator 設定所有的輸入訊號

Waveform1 的 Logic Analyzer 顯示測試 counter 所得的結果，如圖 13-30 所示。

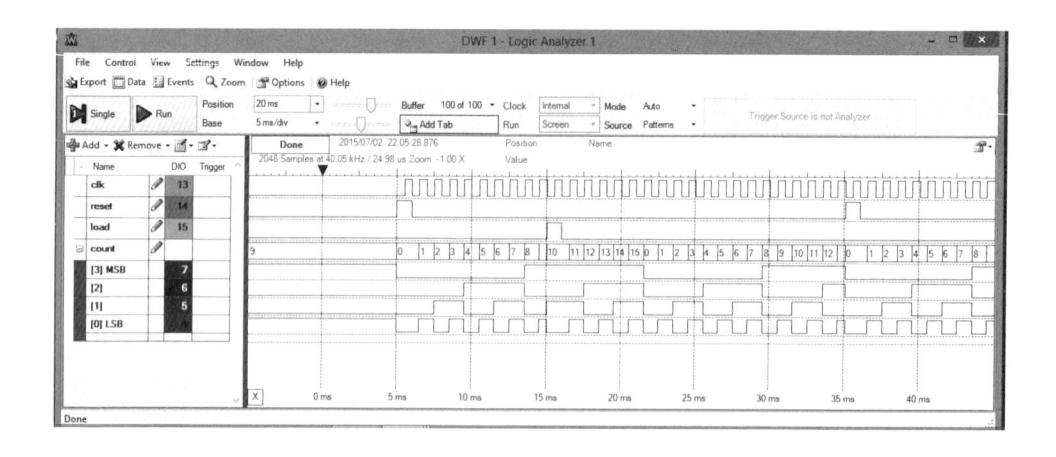

◌ 圖 13-30 ◌　　Logic Analyzer 顯示測試 counter 所得的結果

13-6 課外練習

1. 試用 BASYS3 的 FPGA 來合成第二章的 Simple Adder。並用 Analog Discovery 的 Digital Pattern Generator 和 Logic Analyzer 來測試之。

2. 試用 BASYS3 的 FPGA來合成第三章的 JKFF。並用 Analog Discovery 的 Digital Pattern Generator 和 Logic Analyzer 來測試之。

3. 試用 BASYS3 的 FPGA 來合成第四章的 3 to 8 Decoder。並用 Analog Discovery 的 Digital Pattern Generator 和 Logic Analyzer 來測試之。

4. 試用 BASYS3 的 FPGA 來合成第五章的 Shift Register。並用 Analog Discovery 的 Digital Pattern Generator 和 Logic Analyzer 來測試之。

5. 試述 Program Basys3 板子上的 serial flash 來做 counter 的步驟。

6. 試從圖 13-29、圖 13-30 和 pin.xdc 寫出 Basys3、Analog Discovery、和 Counter 間的接線的關係。

附錄 A　VHDL 電路檔的結構與格式

　　圖 A-1 為 VHDL 電路檔的最基本結構與格式。以一個 and gate 為例，它分成 Entity 和 Architecture 二個部份。Entity 為電路的外部接腳。Architecture 為電路的內部結構。A、B、Y 都代表訊號 Signal，Logic 的訊號最簡單的是 BIT。BIT 只有 '0' 和 '1' 二種**狀態** (State)。

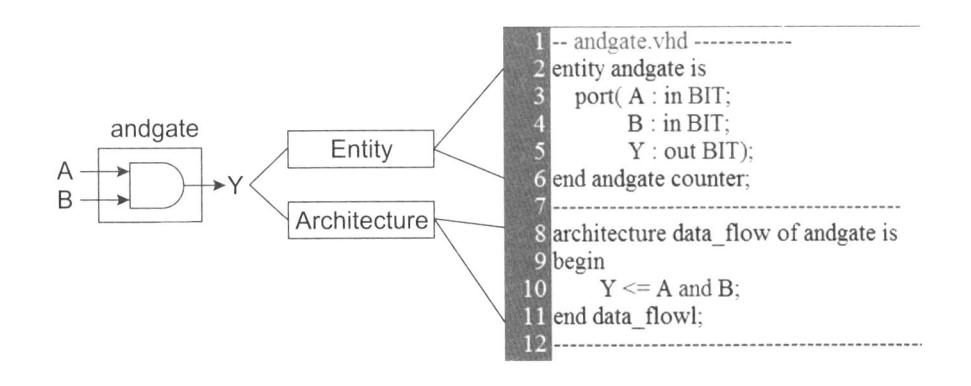

```
1  -- andgate.vhd ------------
2  entity andgate is
3     port( A : in BIT;
4           B : in BIT;
5           Y : out BIT);
6  end andgate counter;
7  ------------------------------
8  architecture data_flow of andgate is
9  begin
10     Y <= A and B;
11 end data_flowl;
12 ------------------------------
```

　　圖 A-1　　VHDL 電路檔的最基本結構與格式

　　比較完善的 IEEE 標準 Logic 訊號是 IEEE.std_logic_1164。它對設計者在除錯上，更有幫助。圖 A-2 便是使用它來代表 Logic Signal 的寫法。擺在 entity 的最前端，宣告電路使用 IEEE Library。

```
1  -- andgate.vhd -----------
2  library IEEE;
3  use IEEE.std_logic_1164.all;
4  ----------------------------------
5  entity andgate is
6    port( A : in std_logic;
7          B : in std_logic;
8          Y : out std_logic);
9  end andgate counter;
10 ----------------------------------
11 architecture data_flow of andgate is
12 begin
13     Y <= A and B;
14 end data_flowl;
15 ----------------------------------
```

圖 A-2　使用標準 IEEE.std_logic_1164 作為訊號的寫法

VHDL 是對電路的一種描述，它跟一般 Program 不同。因為電路的運作，除了先做宣告 Process 後，才會像一般 Program 由 Line 1 按順序一條條 Process 處理下去。否則像圖 A-2 的 VHDL 從 Line 2 到 Line 14 是 concurrent "同時處理"。圖 A-3 的計數器電路 counter，便是一個例子。

```
1  -- 4bit UP counter -----------
2  library IEEE;
3  use IEEE.std_logic_1164.all;
4  IEEE.numeric_std.all;
5  entity counter is
6  generic(n : NATURAL := 4);
7  port(clk : in std_logic;
8       reset : in std_logic;
9       load : in std_logic;
10      Data: in unsigned(n-1 downto 0);
11      count : out std_logic_vector(n-1 downto 0));
12 end entity counter;
13 architecture rtl of counter is
14 begin
15     p0: process (clk, reset, load) is
16        variable cnt : unsigned(n-1 downto 0);
17        begin
18          if reset = '1' then
19             cnt := (others => '0');
20          elsif load = '1' then
21             cnt := Data;
22          elsif rising_edge(clk) then
23             cnt := cnt + 1;
24          end if;
25        count <= std_logic_vector(cnt);
26       end process p0;
27 end architecture rtl;
```

圖 A-3　計數器電路 counter 是 Process 順序處理的例子

圖 A-3 的 VHDL 宣告 p0 是一個 Process，所以從 Line 15~26 是跟一般 Program 相同，由 Line 15 按順序一條條處理到 Line 26 為止。Process 的 (clk, reset, load) 又稱謂 sensitivity list，意思是當 list 內的訊號之一有變化時才進行 Process。

圖 A-3 中，VHDL 的 Line 10 的 Data 和 11 的 count 為 VHDL 對 Bus 寬度的一種描述。而 line 22 的 rising_edge(clk)，說明了當 clk 訊號由 '0' 上升到 '1' 的瞬間，變數 cnt 才從原來的數目上加 1。變數改變的符號用 := ，它代表立即生效。訊號改變的符號用 <= ，它代表要整個 Process 完成後才生效。

附錄 B　VHDL 測試檔的結構與 Stimulus 的寫法

圖 B-1 是一個用來測試 VHDL andgate 電路的 VHDL testbench 測試檔的結構。由於 VHDL andgate 使用 IEEE.std_logic_1164，所以 VHDL testbench 也必須用它。

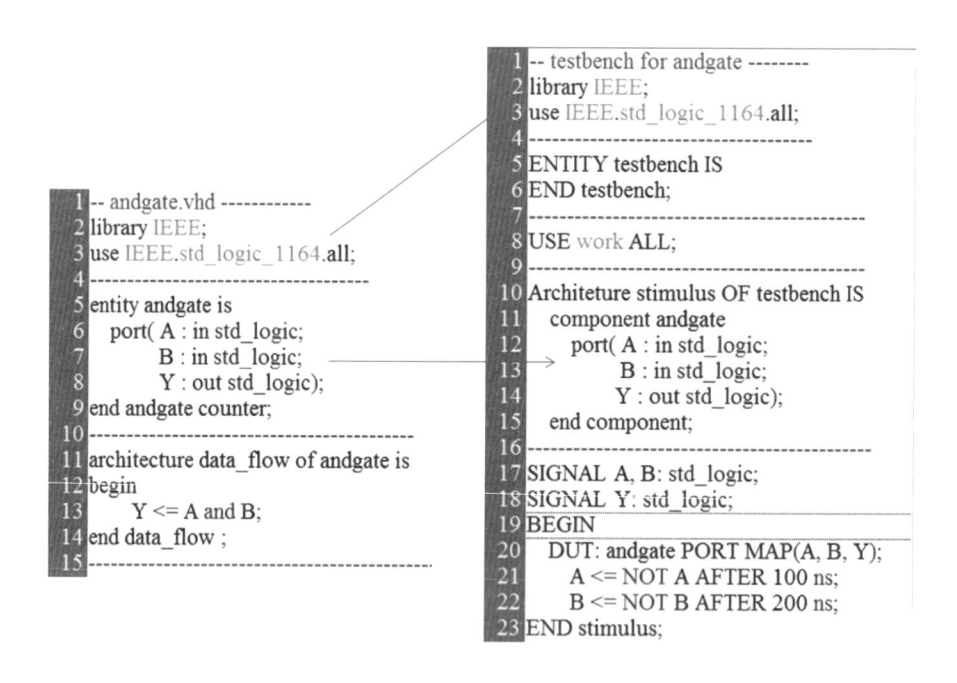

```
1 -- andgate.vhd ------------
2 library IEEE;
3 use IEEE.std_logic_1164.all;
4 ---------------------------------
5 entity andgate is
6   port( A : in std_logic;
7         B : in std_logic;
8         Y : out std_logic);
9 end andgate counter;
10 ---------------------------------
11 architecture data_flow of andgate is
12 begin
13     Y <= A and B;
14 end data_flow ;
15 ---------------------------------
```

```
1 -- testbench for andgate --------
2 library IEEE;
3 use IEEE.std_logic_1164.all;
4 ---------------------------------
5 ENTITY testbench IS
6 END testbench;
7 ---------------------------------
8 USE work ALL;
9 ---------------------------------
10 Architeture stimulus OF testbench IS
11   component andgate
12     port( A : in std_logic;
13           B : in std_logic;
14           Y : out std_logic);
15   end component;
16 ---------------------------------
17 SIGNAL A, B: std_logic;
18 SIGNAL Y: std_logic;
19 BEGIN
20   DUT: andgate PORT MAP(A, B, Y);
21   A <= NOT A AFTER 100 ns;
22   B <= NOT B AFTER 200 ns;
23 END stimulus;
```

圖 B-1　VHDL testbench 測試檔的組成

VHDL testbench 測試檔也是由 ENTITY 和 Architecture 二個部份所組成。它和 VHDL 電路檔不同的是 testbench 的 ENTITY 內沒有 I/O Port。它的 Architecture 的首項是被測試的電路 component。

圖 B-1 中的 testbench Line 11~15 就是從 andgate.vhd 的 Line 5~9 搬了過

200

來，把其中的 entity 改為 component 而成。

Testbench 的功能是產生對 VHDL 電路的輸入訊號，在圖 B-1 的例子裡，Line 17~18 採用和 andgate.vhd 完全相同的訊號名稱 A、B，和 Y。這樣的做法，可以讓 Line 20 最簡化。

Line 21~22 是一種產生 "連續對稱型方波" 的寫法。方波 A 的週期為 200 ns，頻率為 5E6 Hz，方波 B 的週期為 400 ns，頻率為 2.5E6 Hz。

B.1 如何產生各種不同的 Stimulus 訊號

書寫 Testbench.vhd 最重要的部份，是產生 Stimulus 輸入信號。

歸納起來，輸入信號一共有以下的五種：

1. 對稱並且重複的波形，Clock 就是它的代表，它的寫法如下：

   Signal clock：bit := '1';

   clock <= NOT clock AFTER 30 ns;

   也可以寫成：

   Signal clock：bit := '1';

   Wait for 30 ns;   clock <= NOT clock;

2. 單一的波形，reset 就是它的代表，它的寫法如下：

   Signal reset：bit := '0';

   Wait for 30 ns;   reset <= '1';

   Wait for 30 ns;   reset <= '0';

   Wait;

3. 不規則，而且不重複，多個 bit 的波形。它的寫法如下：

   Constant wave: bit_vector(1 to 8):= "10110100";

   Signal x: bit := '0';

   For i IN wave'RANGE LOOP

   x <= wave(i); wait for 30 ns;

   END LOOP;

   Wait;

4. 不規則，但重複，多個 bit 的波形。它的寫法如下：

   Constant wave: bit_vector(1 to 8):= "10110100";

   Signal y: bit := '0';

   For i IN wave'RANGE LOOP

   y <= wave(i); wait for 30 ns;

   END LOOP;

5. 數字型，多個 bit 的波形。它的寫法如下：

   Signal z: INTEGER RANGE 0 to 255;

   Z <= 0; wait for 120 ns;

   Z <= 33; wait for 120 ns;

   Z <= 255; wait for 60 ns;

   Z <= 99; wait for 180 ns;

B.2 產生五種不同的 Stimulus 訊號的 VHDL testbench

產生以上五種不同 VHDL test signals 的 VHDL testbench，如圖 B-2 所示。

```vhdl
Library ieee;
Use ieee.std_logic_1164.all;
-------------------------------------------
Entity test_testbench is
End test_testbench;
-------------------------------------------
Architecture testbench of test_testbench is
    Signal a: std_logic := '1';
    Signal b: std_logic := '0';
    constant wave: bit_vector(1 to 8):="10110100";
    Signal x: bit := '0';
    Signal y: bit := '0';
    Signal z: integer range 0 to 255;
Begin
    ----- Generate a: -----
    Process
    Begin
        wait for 30 ns;
        a <= not a;
    End Process;
    ----- Generate b: -----
    Process
    Begin
        wait for 30 ns;
        b <= '1';
        wait for 30 ns;
        b <= '0';
        wait;
    End Process;
    ----- Generate c: -----
    Process
    Begin
        For i in wave'range loop
            x <= wave(i); wait for 30 ns;
        end loop;
        wait;
    End Process;
    ----- Generate d: -----
    Process
    Begin
        For i in wave'range loop
            y <= wave(i); wait for 30 ns;
        end loop;
    End Process;
    ----- Generate e: -----
    Process
    Begin
        z <= 0;    wait for 120 ns;
        z <= 33;   wait for 120 ns;
        z <= 255;  wait for  60 ns;
        z <= 99;   wait for 180 ns;
    End Process;
-------------------------------------------
End testbench;
```

圖 B-2　　產生五種代表波形的 VHDL 寫法

如果用 ModelSim 來做 Simulation，可獲得的波形，如圖 B-3 所示。

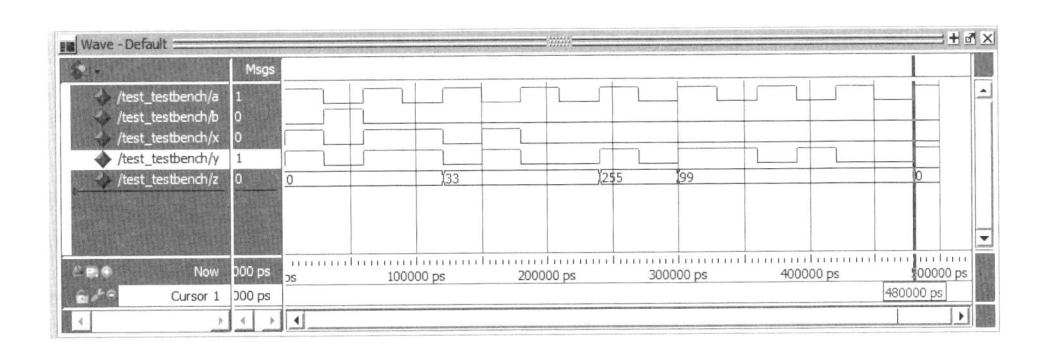

圖 B-3　　ModelSim simulator 顯示的五種不同波形

附錄 C 如何使用和產生 Digital Pattern

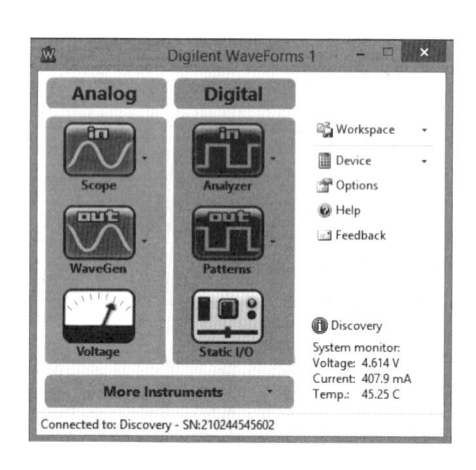

C.1 Digital Pattern Generator 的設定

點擊 **Digital** 部份的 **out**，當 Digital pattern Generator 出現時，首先要清除舊有的設定。點擊 **Remove>>Clear List**，結果如圖 C-1 所示。

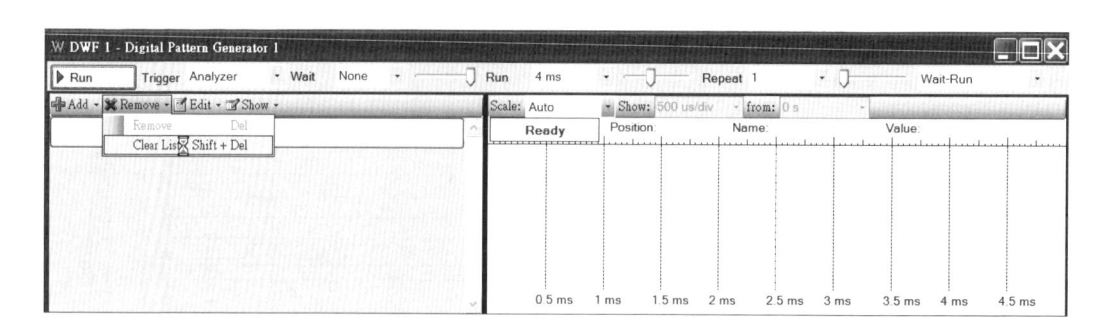

圖 C-1　清除數字模式產生器舊有的設定

接下來是要加入 (+Add) 什麼？Signals、Bus、還是 Empty Bus。由於這是一個簡單的例子，選用 **Signals** 即可，如圖 C-2 所示。

204

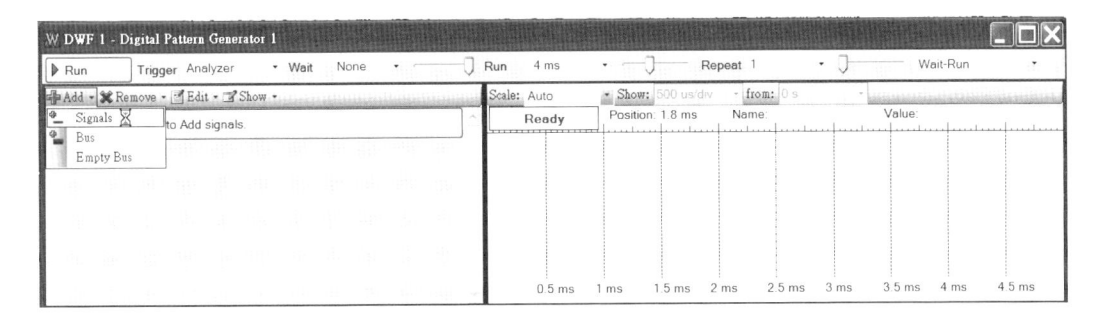

附錄 C　如何使用和產生 Digital Pattern　205

圖 C-2　　選用加入 Signals

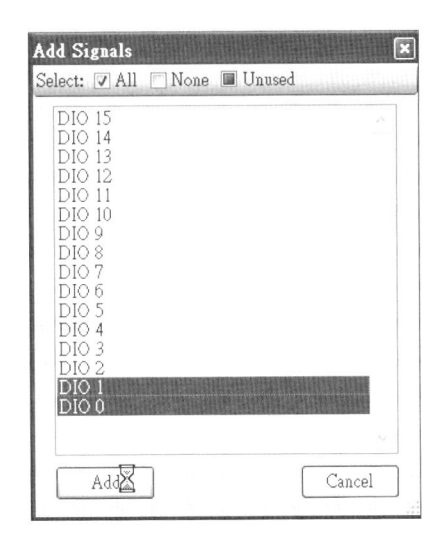

圖 C-3　　Add signals 視窗的出現

選用 **Signals** 後便有圖 C-3 Add signals 視窗的出現。

這個例子的輸入訊號只有 A 和 B 二個，所以只要在 DIO 0~DIO 16 中任選 2 個就夠了。這裡選用的是 DIO 0 和 DIO 1。

DIO 的 Type 共 4 種，分別為 Constant、Clock、Random 和 Custom，如圖 C-4 所示。

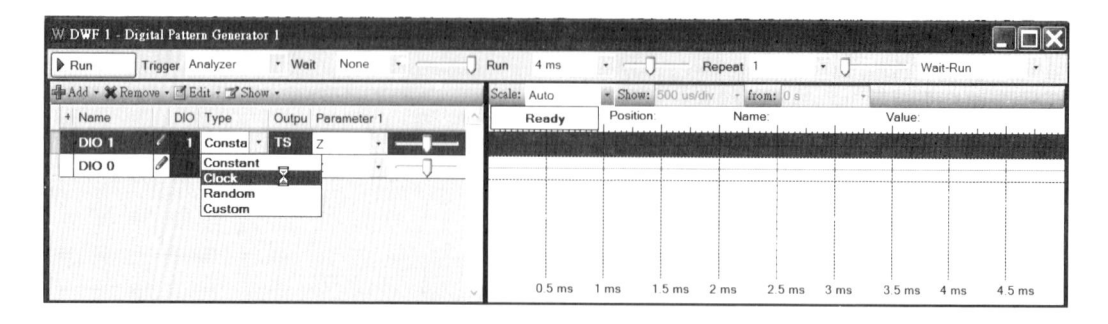

◎ 圖 C-4 ◎　　　　DIO 的 Type 的選擇和設定

　　這裡選用的是 Clock。DIO 0 的輸出為 500 Hz，DIO 1 的輸出為 1000 Hz。這樣的結果，恰好構成 Truth Table 中的 00、01、10 和 11 的 2-input Binary 信號的組合。結果如圖 C-5 所示。

　　其中有一項 Trigger，要特別注意，它必須選定為 Analyzer，那是說如果 Analyzer 不啟動，DIO 0 和 DIO 1 不輸出。

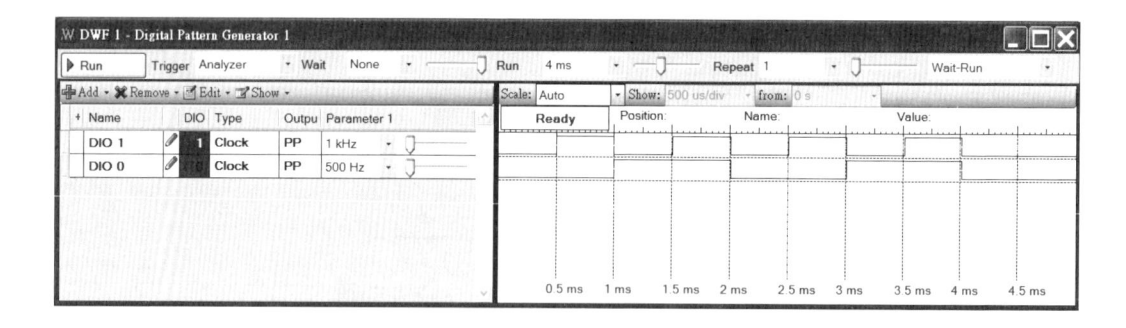

◎ 圖 C-5 ◎　　　　DIO 0 和 DIO 1 的輸出信號

附錄 C　如何使用和產生 Digital Pattern　207

　　Digital Patterns 也可以用 Bus 來產生，如圖 C-6 所示。從選取 +Add >> Bus 開始，再將 DIO 0 和 DIO 1 加入到 Bus 0。Format 選用 Binary，LSB 為 DIO 0，MSB 為 DIO 1。

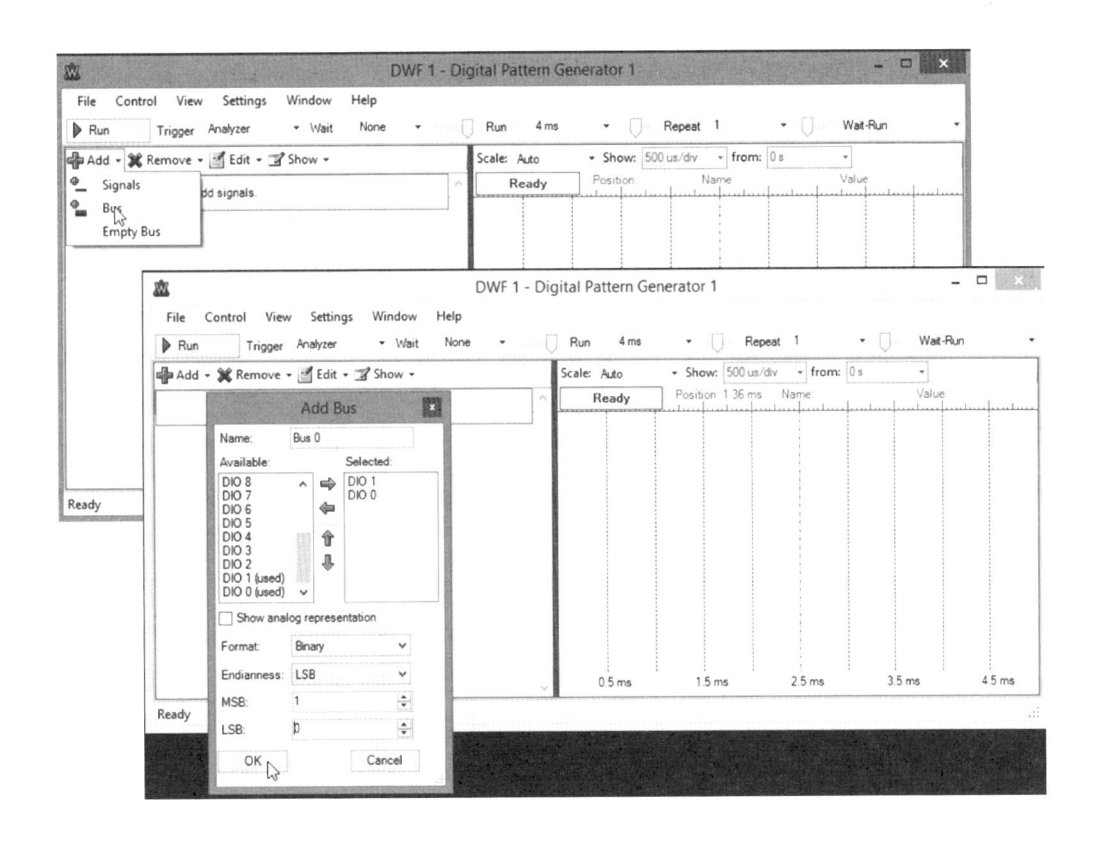

圖 C-6　　Digital Pattern Generator 選用 Binary Counter Bus 的設定之一

可以供 Bus 設定的 Patterns 有 10 種之多，如圖 C-7 所示。如果是選用 Binary Counter，只要單擊它一下便成了。

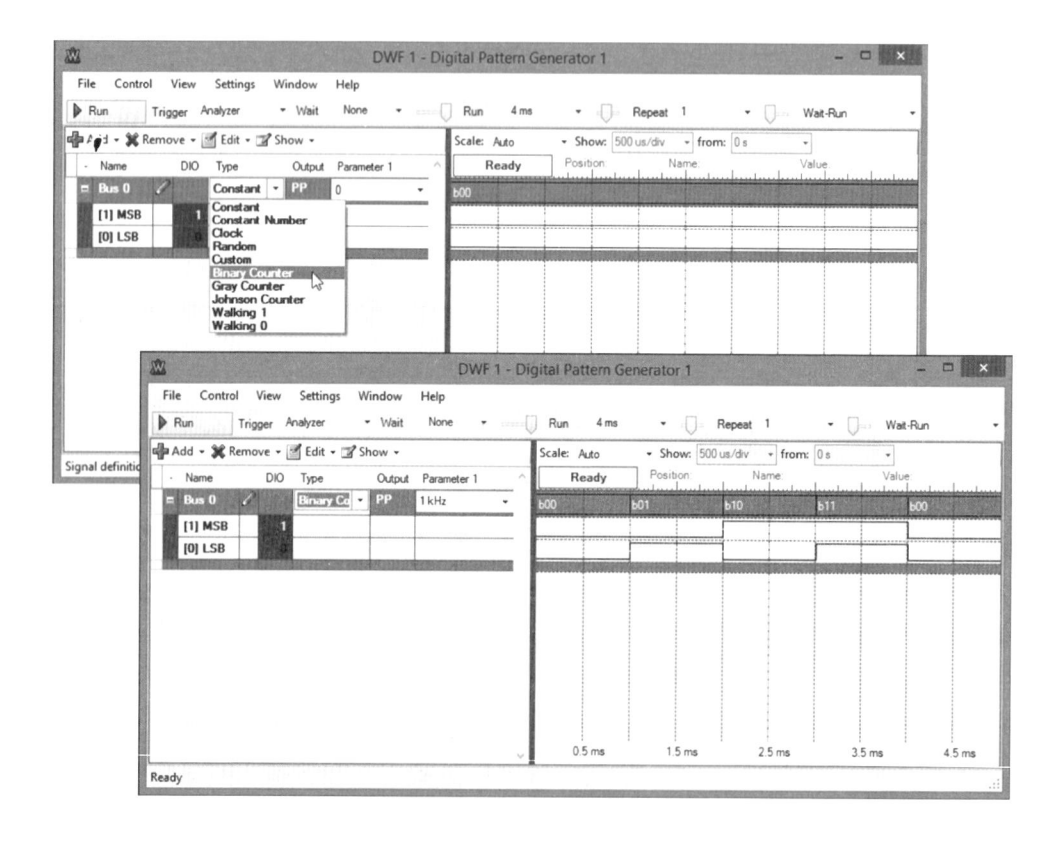

圖 C-7　　Digital Pattern Generator 選用 Binary CounterBus 的設定之二

附錄 C　如何使用和產生 Digital Pattern　209

　　如果 Bus 的 Patterns 不在所提供的範圍內，可以選用 Custom，如圖 C-8 所示。然後選用 Edit Parameters of "Bus 0"。

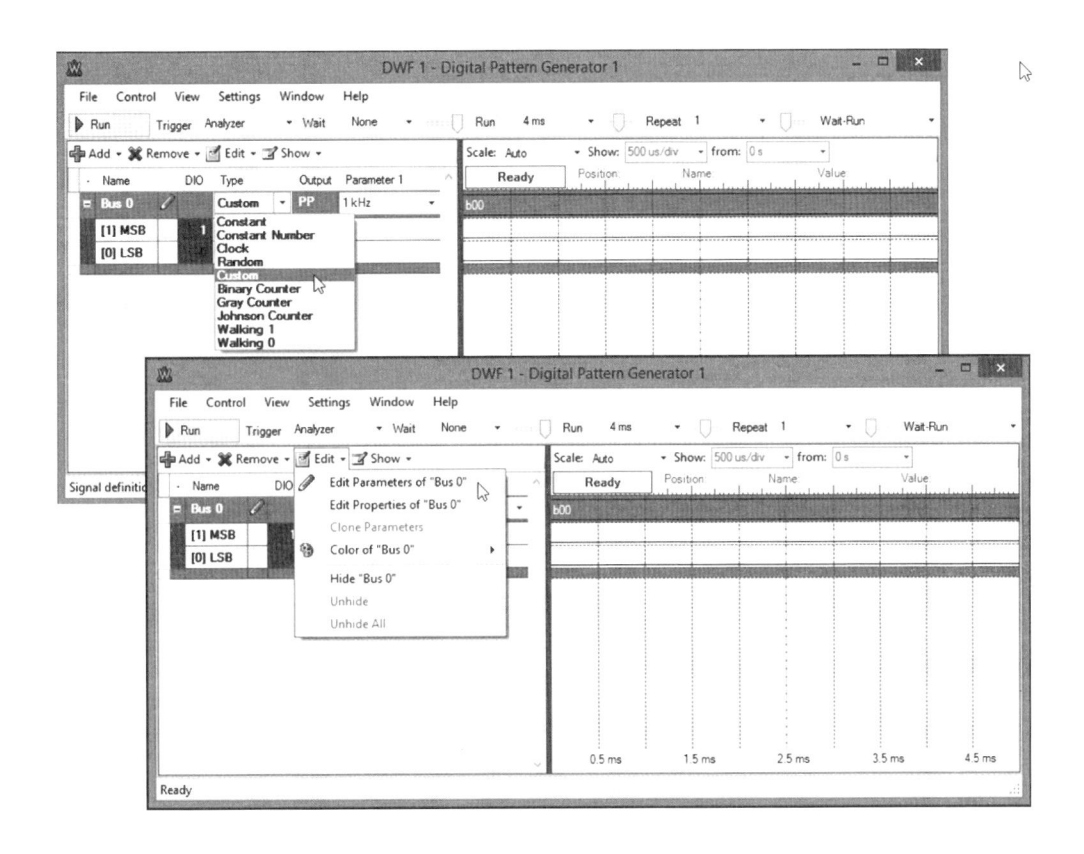

圖 C-8　Digital Pattern Generator 選用 Customer Bus 的設定之一

Customer Bus 的設定之二，首先要設定 Digital Pattern 的各個參數，其中有 Type、Output、Idle、Buffer Size、Frequency 等項。接下來為波形的設定，把 Mouse 的指針移到 LSB 或 MSB 的位置，點擊 '0'，'1'，或 'Z' 以決定在該時間內 Logic 的性質。最後還須點擊圖 C-9 右上方的 Lock，以固定所有的設定。設定的結果如圖 C-10 所示。

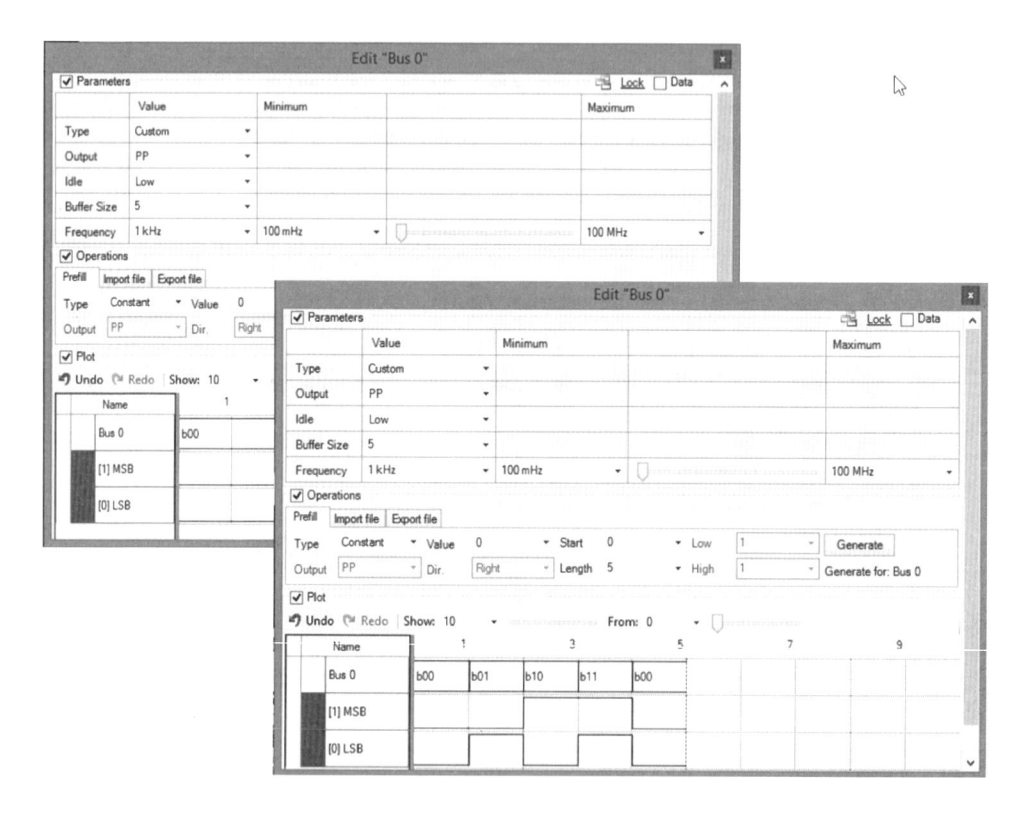

圖 C-9　Digital Pattern Generator 選用 Customer Bus 的設定之二

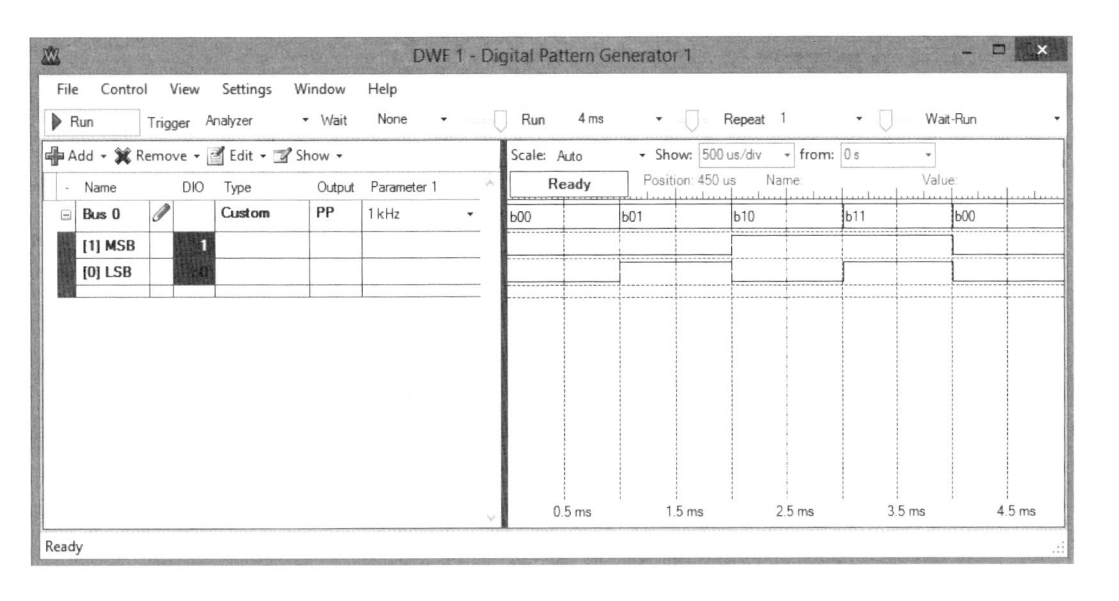

附錄 C　如何使用和產生 Digital Pattern　**211**

圖 C-10　Customer Bus 的設定的結果

附錄 D　如何使用 Logic Analyzer

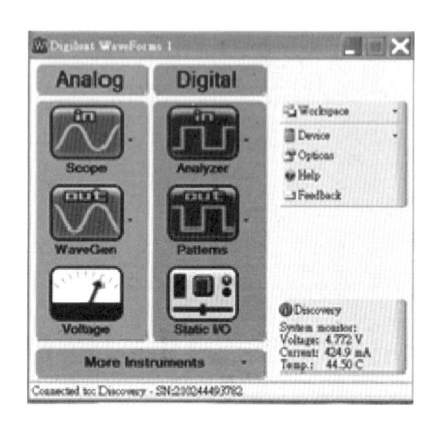

點擊 **Digital** 部份的 **in**，當 Logic Analyzer 出現時，首先也是要清除舊有的設定。點擊 **Remove>>Clear List**，結果如圖 D-1 所示。

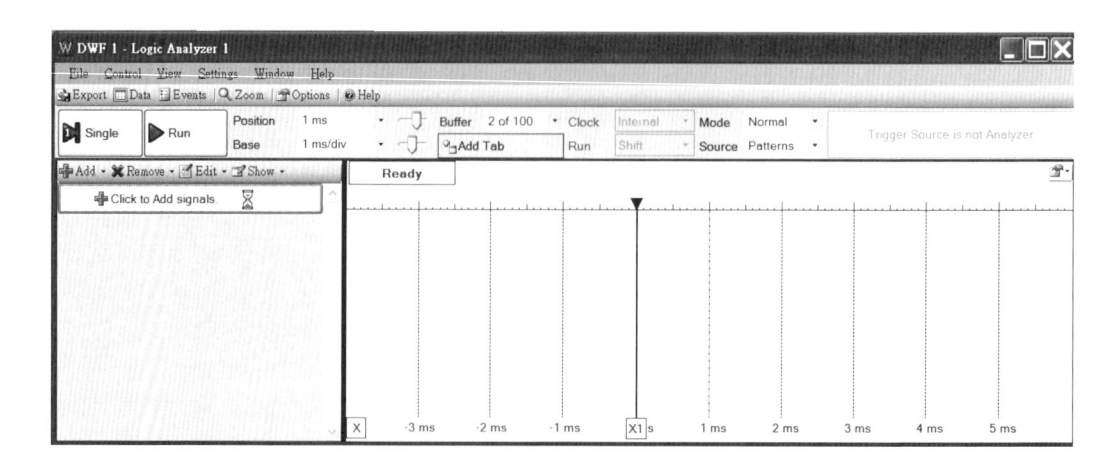

圖 D-1　清除 Logic Analyzer 舊有的設定

212

附錄 D　如何使用 Logic Analyzer　213

　　用第一章 p.6 圖 1-9 為例它的輸出訊號有 C、D、E 和 F 等 4 個，照理只要在選用 DIO 2~DIO 16 中任選 4 個就夠了。但是為了能完成整個系統的 Timing Diagram，可以將輸入的 DIO 0 和 DIO1 包括在內，因此選用了 DIO 0~DIO5 等 6 個訊號。如圖 D-2 所示。

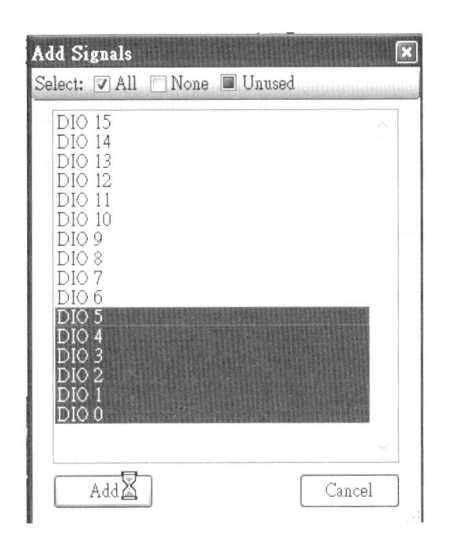

圖 D-2：　　為了 Timing Diagram 選用 DIO 0~DIO5 等 6 個訊號

設定的結果，如圖 D-3 所示。此時 Logic Analyzer 在 Ready 狀態，只要 Pattern Generator 一啓動，它馬上記錄電路輸入 A、B 和輸出 C、D、E、F 各個訊號的 Timing Diagram。

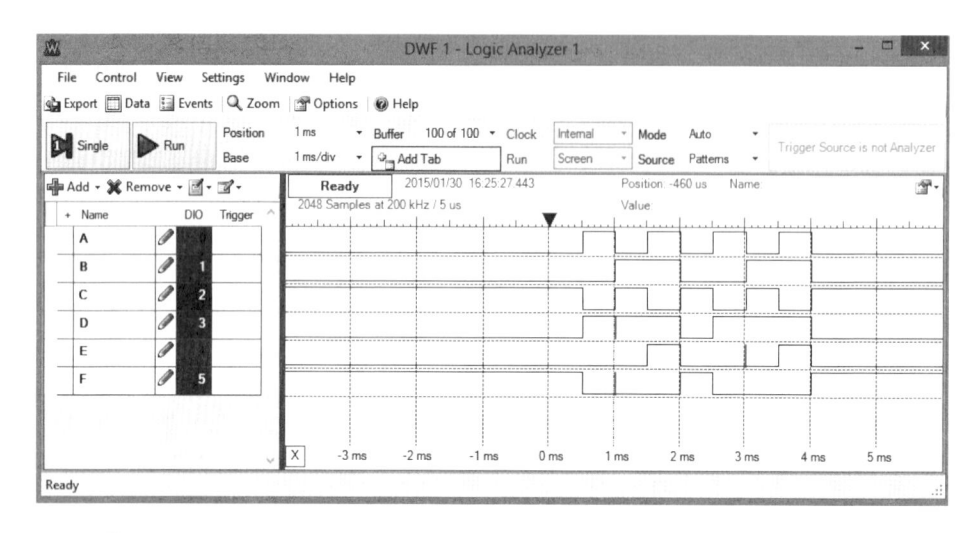

☙ 圖 D-3 ☙　　　Logic Analyzer 在 Ready 等待 Pattern Generator 的啓動

附錄 E　ModelSim Simulation 模擬測試

使用 ModelSim Simulator 來對所設計的 VHDL 電路做模擬測試，可以分成以下的 4 個階段：

1. 第一個階段：將所設計的 VHDL 電路和它的 TestBench 測試檔，放在同一個檔案夾內，然後開啟 ModelSim，如圖 E-1 所示。

圖 E-1　SN74LS00.vhd 和 Testbench.vhd 同在檔案夾 SN74LS00 內

待 ModelSim Welcome 視窗出現後選擇 Jampstart，如圖 E-2 所示。

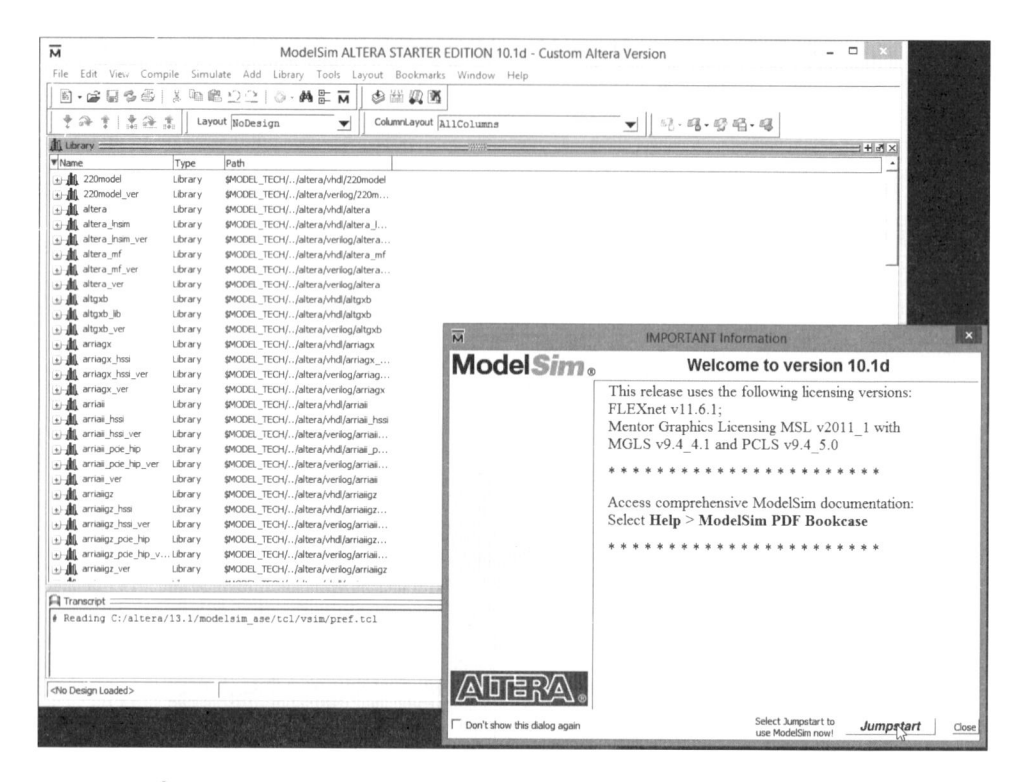

☆ 圖 E-2 ☆　　　　在 ModelSim Welcome 視窗出現後選擇 Jampstart

接下來有一連串的選用如 Create a Project、設定 SN74LS00 檔案夾的位置、和加入所需的 vhd files 等，如圖 E-3 和圖 E-4 所示。

圖 E-3　Create a Project 和檔案夾的位置的選定

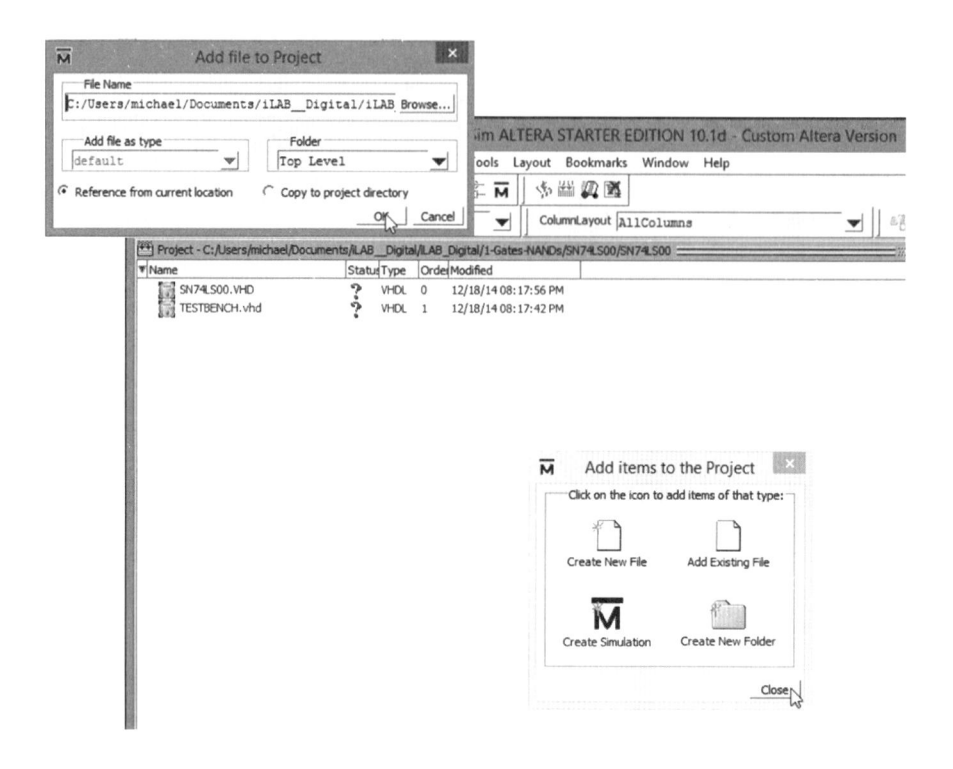

218 iLAB Digital 數位電路設計、模擬測試與硬體除錯

圖 E-4　　　Create a Project 和 SN74LS00.vhd 及 Testbench.vhd 的選定

附錄 E　　ModelSim Simulation 模擬測試　**219**

2. 第二個階段：**編輯** (Compile)，檢查所參與的 files 是否合乎規則？如圖 E-5 所示。

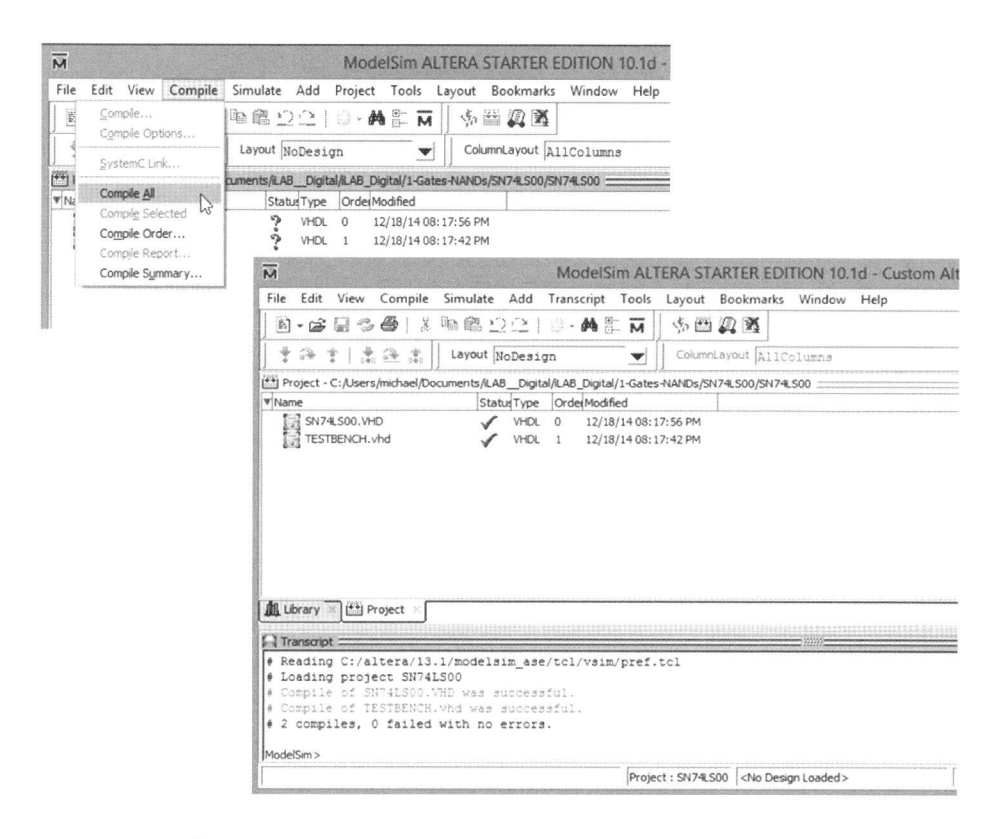

　圖 E-5　　　　通過 Compile 所有的 vhd files 都合乎規則

3. 第三個階段：是模擬測試 Simulation，在 Start simulation 視窗的 work 裡選取 testbench，然後 OK，如圖 E-6 所示。

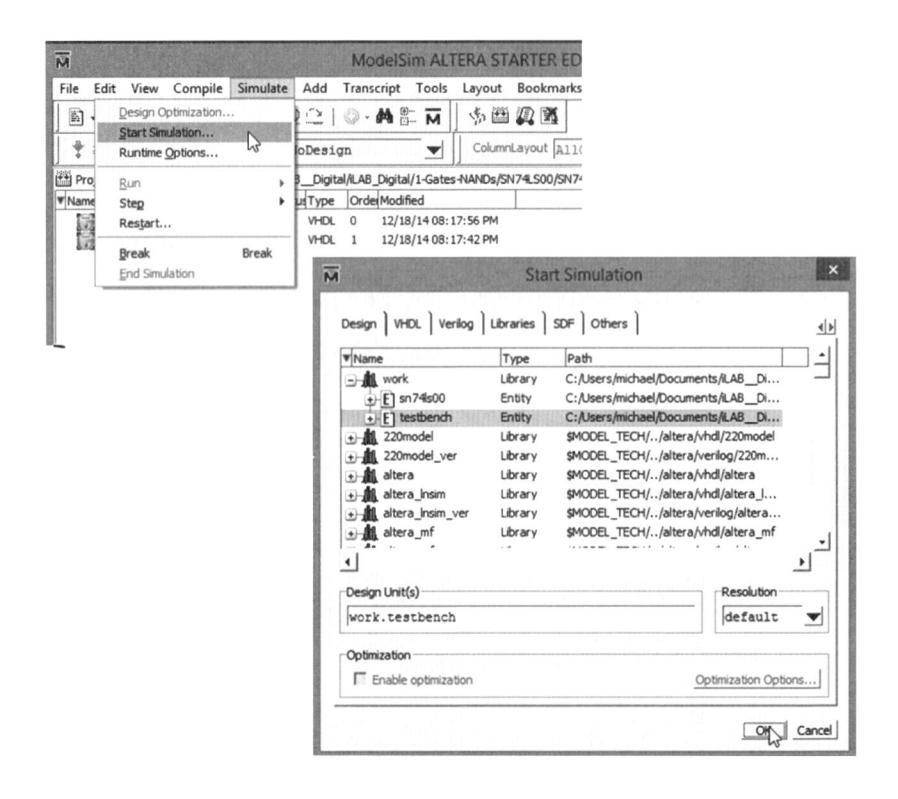

◌ 圖 E-6 ◌　　　在 Start simulation 視窗的 work 中選取 testbench，然後 OK

附錄 E　ModelSim Simulation 模擬測試　**221**

4. 第四個階段：是模擬測試的波形顯示。右擊 Objects 視窗的空白處，依次選用 Add to>>Wave>>Signal in Region，如圖 E-7 所示。

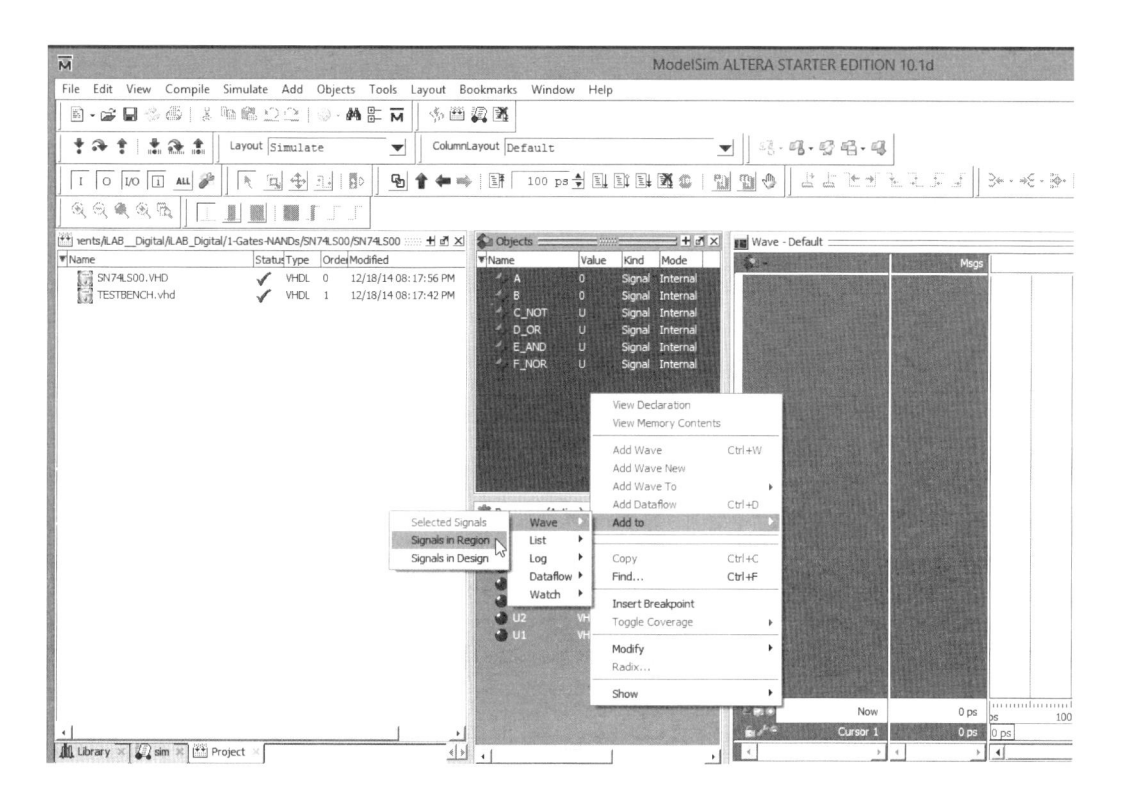

圖 E-7　右擊 Objects 視窗的空白處，依次選用 Add to>>Wave>>Signal in Region

以上的動作是把 Objects window 中的 Objects 搬進 Waveform windows，用來顯示波形，如圖 E-8 所示。

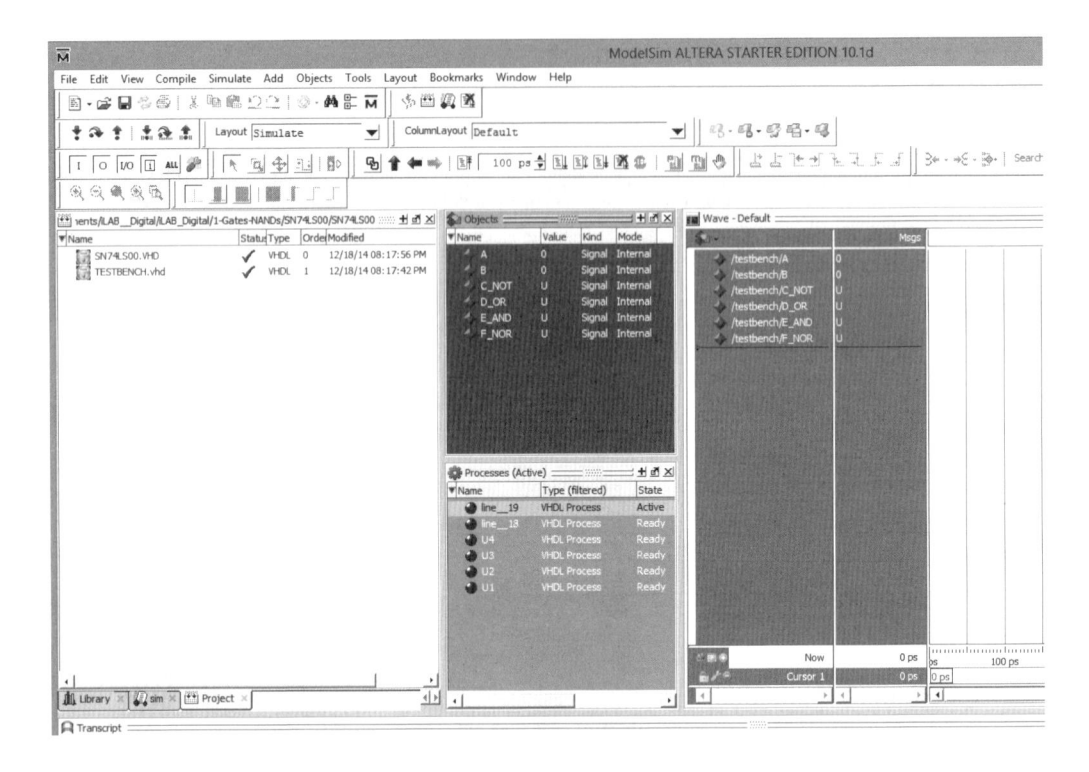

◎ 圖 E-8　　　Objects 搬進 Waveform windows 以便顯示波形

附錄 E　ModelSim Simulation 模擬測試

波形的顯示跟時間的長短有關，在 Testbench 裡，Signal B 的週期最長為 400 ns，所以在 Transcript 視窗令 VSIM > run 400 ns，如圖 E-9 所示。

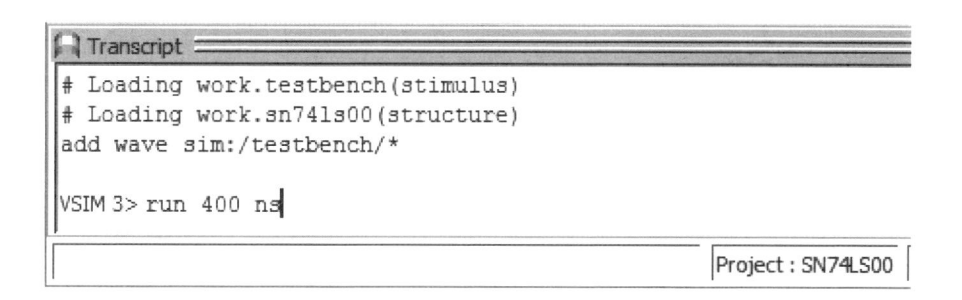

圖 E-9　Transcript 視窗令 VSIM > run 400 ns

當 Wave window 顯示部份波形後，要顯示全部的波形，在 Wave 的選項中選用 Zoom>>Zoom Full，如圖 E-10 所示。圖 E-11 便是它的全部 I/O 測試所得的波形。

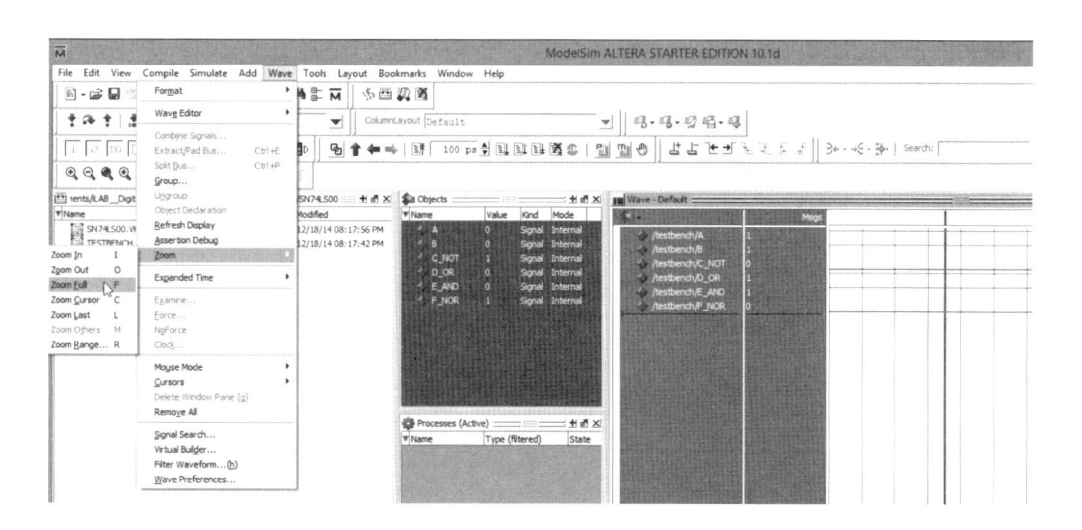

圖 E-10　在 Wave 的選項中選用 Zoom>>Zoom Full 以觀測全部的波形

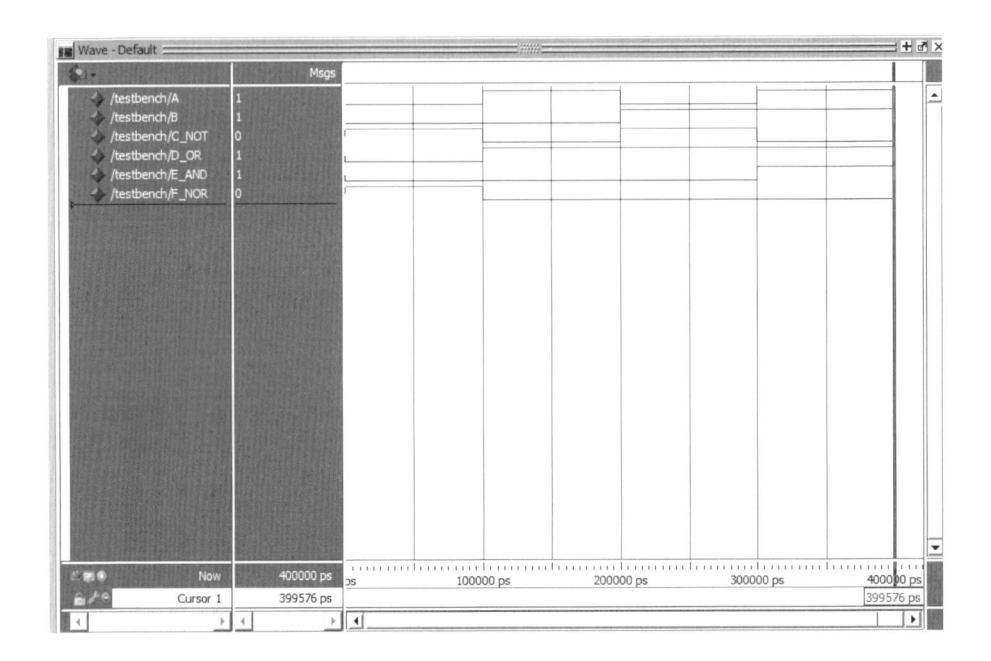

🔾 圖 E-11 🔾　　SN74LS00.vhd 的全部 I/O 測試波形

附錄 F　LM565 PLL 的設定

　　鎖相環積體電路 (Phase Locked Loop IC) 是數位電路不可缺少的一種，也是數位和線性雙重的混合產物。LM565 可用於 500 KHz 以下的電路，它的組成如圖 F-1 所示。

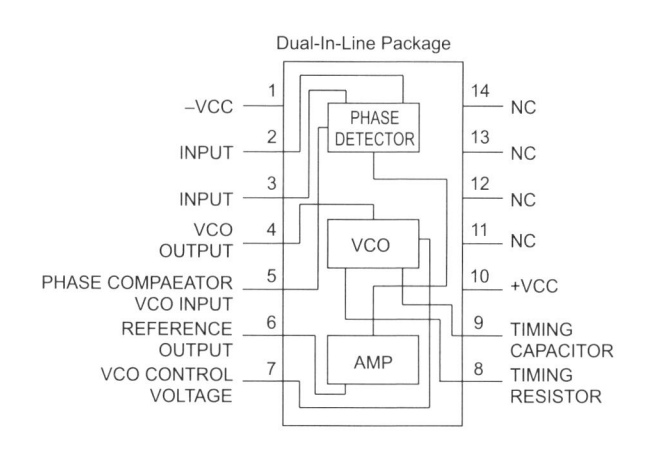

```
Dual-In-Line Package

              1                14
    -VCC  ────┤  PHASE      ├──── NC
              2  DETECTOR   13
    INPUT ────┤             ├──── NC
              3              12
    INPUT ────┤             ├──── NC
    VCO       4              11
    OUTPUT────┤    VCO      ├──── NC
    PHASE     5              10
    COMPAEATOR┤             ├──── +VCC
    VCO INPUT
    REFERENCE 6               9   TIMING
    OUTPUT ───┤             ├──── CAPACITOR
    VCO       7   AMP        8   TIMING
    CONTROL───┤             ├──── RESISTOR
    VOLTAGE
```

Free running frequency Setup:
　Pin 9 to Ground Timing Capacitor
　Pin 8 to Vcc Timing Resistor
Low pass Filter Setup:
　Pin 7 to Vcc Filter Capacitor

圖 F-1　LM 565 PLL 的組成

電路的使用連線，和必要的 RC 附件，如圖 F-2 所示。

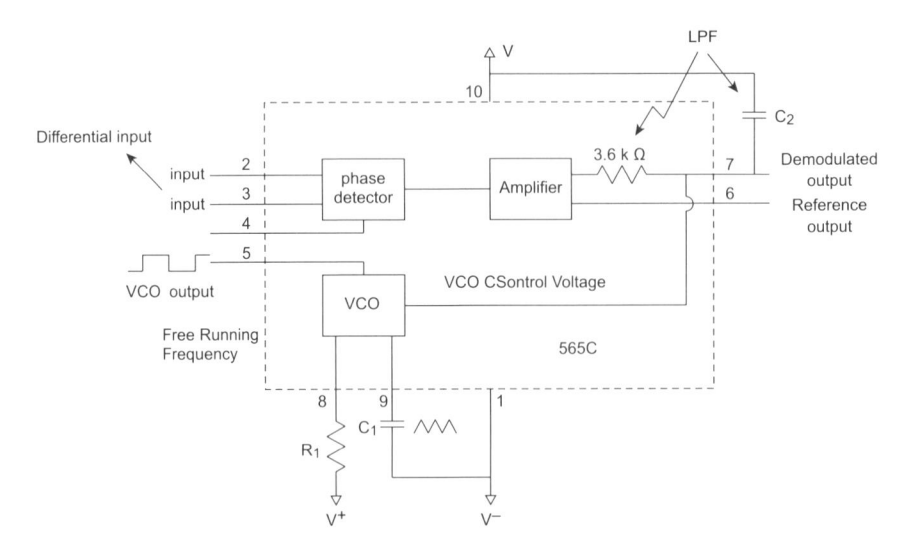

圖 F-2 LM565 電路的使用連線，和必要的 RC 附件

它的 free-running frequency、lock range 和 capture range 和所採用的電阻 R、電容 C 之間的關係，如圖 F-3 所列公式所示。

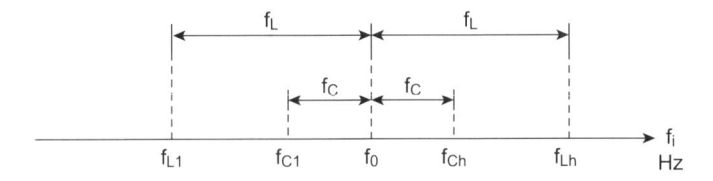

The VOC free-running frequency, f_0

$$f_0 \approx \frac{0.3}{R_1 C_1} \quad (1)$$

The hold-in, tracking, or acquisition range, f_H

$$f_C \approx \frac{\pm 8 f_0}{V^+ + |V^-|} \quad (2)$$

The capture, pull-in, or acquisition range, f_C

$$f_C \approx \pm \sqrt{f_H \cdot f_{lpf}} \quad (3)$$

where, f_{lpf}, is the 3 dB frequency of the lowpass filter section.

圖 F-3 free-running、lock range 和 capture range 與 RC 的關係

依據以上三個公式可以用 C/C++ Program 來計算，如圖 F-4 所示。結果對零件的選擇和採用有很多的幫助。注意 R1 之電阻值，應在 2 k～20 k 之間。

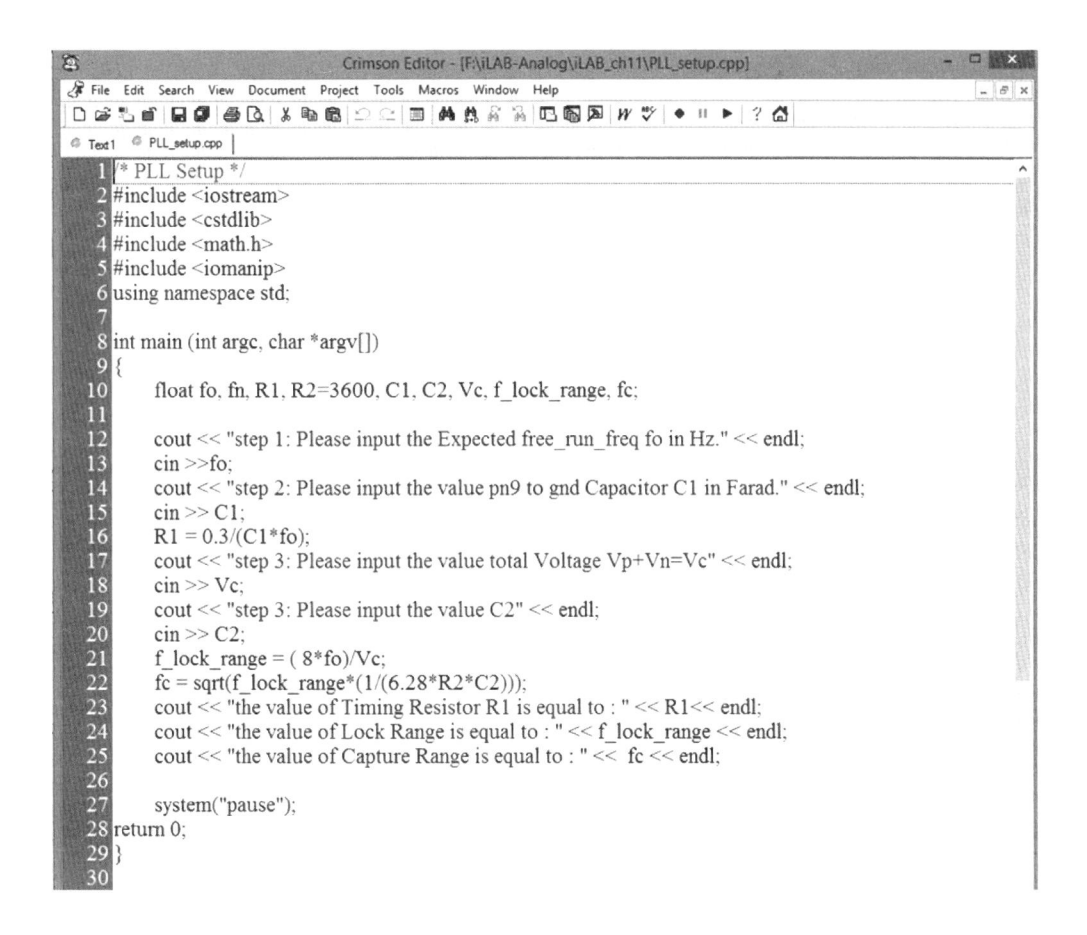

```cpp
/* PLL Setup */
#include <iostream>
#include <cstdlib>
#include <math.h>
#include <iomanip>
using namespace std;

int main (int argc, char *argv[])
{
    float fo, fn, R1, R2=3600, C1, C2, Vc, f_lock_range, fc;

    cout << "step 1: Please input the Expected free_run_freq fo in Hz." << endl;
    cin >>fo;
    cout << "step 2: Please input the value pn9 to gnd Capacitor C1 in Farad." << endl;
    cin >> C1;
    R1 = 0.3/(C1*fo);
    cout << "step 3: Please input the value total Voltage Vp+Vn=Vc" << endl;
    cin >> Vc;
    cout << "step 3: Please input the value C2" << endl;
    cin >> C2;
    f_lock_range = ( 8*fo)/Vc;
    fc = sqrt(f_lock_range*(1/(6.28*R2*C2)));
    cout << "the value of Timing Resistor R1 is equal to : " << R1<< endl;
    cout << "the value of Lock Range is equal to : " << f_lock_range << endl;
    cout << "the value of Capture Range is equal to : " <<  fc << endl;

    system("pause");
return 0;
}
```

圖 F-4　用 C/C++ Program 來計算和選擇所需的 RC 附件

圖 F-5 是對於 free-running frequency 為 100 KHz 時，C1 為 500 pF，Vcc 為 10V，C2 為 0.1 μF 時，所獲得應有的 R1 值和電路可能有的頻率 lock range 和 capture range。

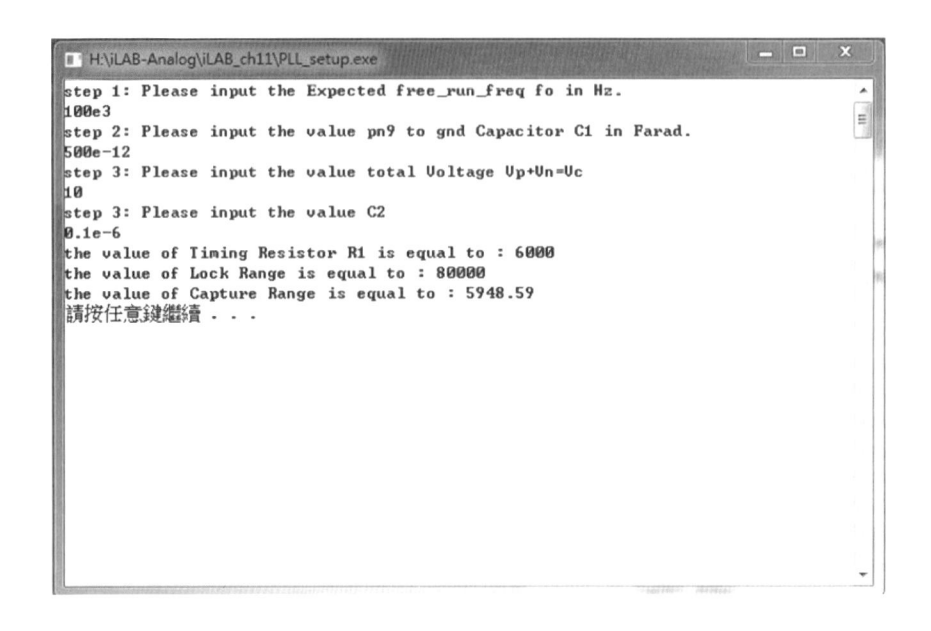

```
H:\iLAB-Analog\iLAB_ch11\PLL_setup.exe

step 1: Please input the Expected free_run_freq fo in Hz.
100e3
step 2: Please input the value pn9 to gnd Capacitor C1 in Farad.
500e-12
step 3: Please input the value total Voltage Up+Un=Uc
10
step 3: Please input the value C2
0.1e-6
the value of Timing Resistor R1 is equal to : 6000
the value of Lock Range is equal to : 80000
the value of Capture Range is equal to : 5948.59
請按任意鍵繼續 . . .
```

圖 F-5　C/C++Program 輸入和輸出的結果。

附錄 G　Xilinx CPLD 的電路合成

　　FPGA 和 CPLD 同屬 Programmable Logic，在性質上類似 RAM 和 EEPROM。FPGA 在 Program 完成之後，不可以 Power Off，否則又須重新 Program。CPLD 不同的地方是當 Program 完成之後，不受 Power Off 的影響。除非要更改 Program 的內容，才需要重新 Program。至於用來 Program 的軟體幾乎是完全相同，Altera 用的是 Quartus II，Xilinx 用的是 ISE，各用各的軟體來 Program 各自的產品，雖然內容相仿，無法相調使用。

G.1　Xilinx 的 CoolRunner-II CPLD Starter Kit

　　為了推廣 CPLD，Xilinx 推出了 CoolRunner-II CPLD Starter Kit，如圖 G-1 所示。

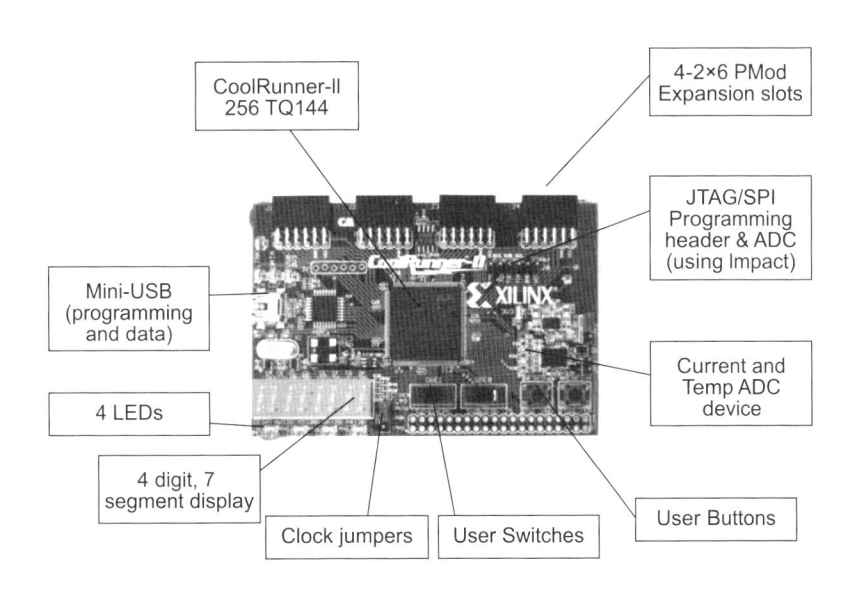

圖 G-1　CoolRunner-II CPLD Starter Kit 的電路板

Program CoolRunner-II 所用的軟體為 ISE 11.1 webpack。由該軟體處理後所得到的 .JED file 可以用 USB-Mini Cable 接到 PC，經過 CoolRunner-II Utility Window 直接 Program CoolRunner-II PLD，如圖 G-2 所示。

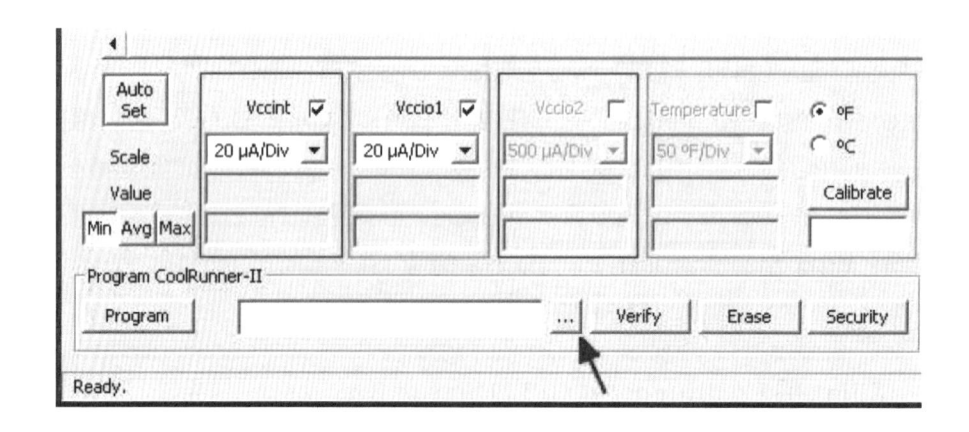

圖 G-2　　CoolRunner-II Utility Window 直接 Program

這個附錄將示範如何將一個 4 bits 的 counter 經由 Xilinx ISE11.1 獲得 Program CoolRunner-II 256 TQ144 CPLD 的 Counter.jed 檔案。再經由 CoolRunner-II Utility Window 從 PC 經 USB-mni cable 完成 Program 的每一個步驟。

G.2　ISE 11.1 webpack 處理 4 bit Counter 的步驟

1. 第一個階段：是在 ISE Project Navigator 下，決定要處理的檔案和 CPLD 的名稱，如圖 G-3 所示。

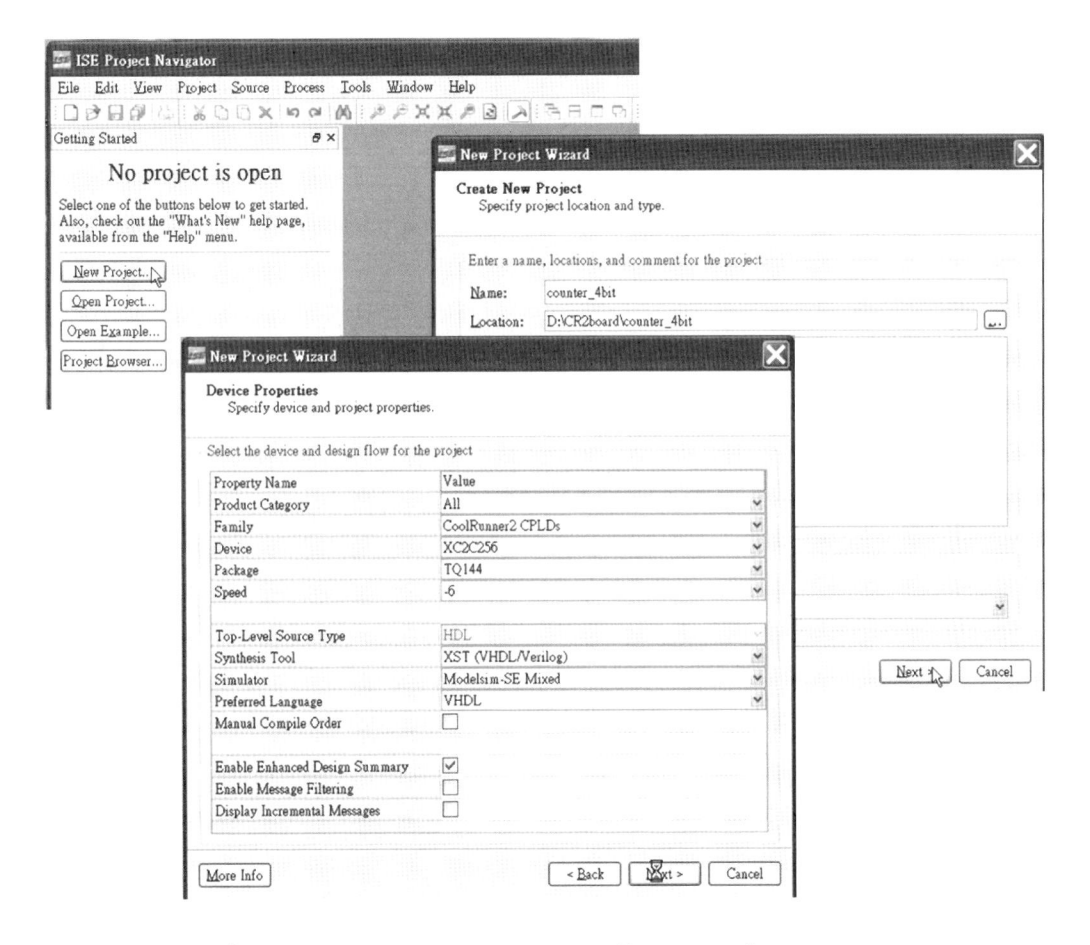

圖 G-3　檔案 counter_4bit 使用 CPLD 為 CXC2C256

接下來是使用 New Project Wizard，決定是否要 Create New Source 或 Add existing Source？由於 Counter_4bit 是一個簡單的設計，所以全部不需要而選用 Next。最後如圖 G-4 所示。

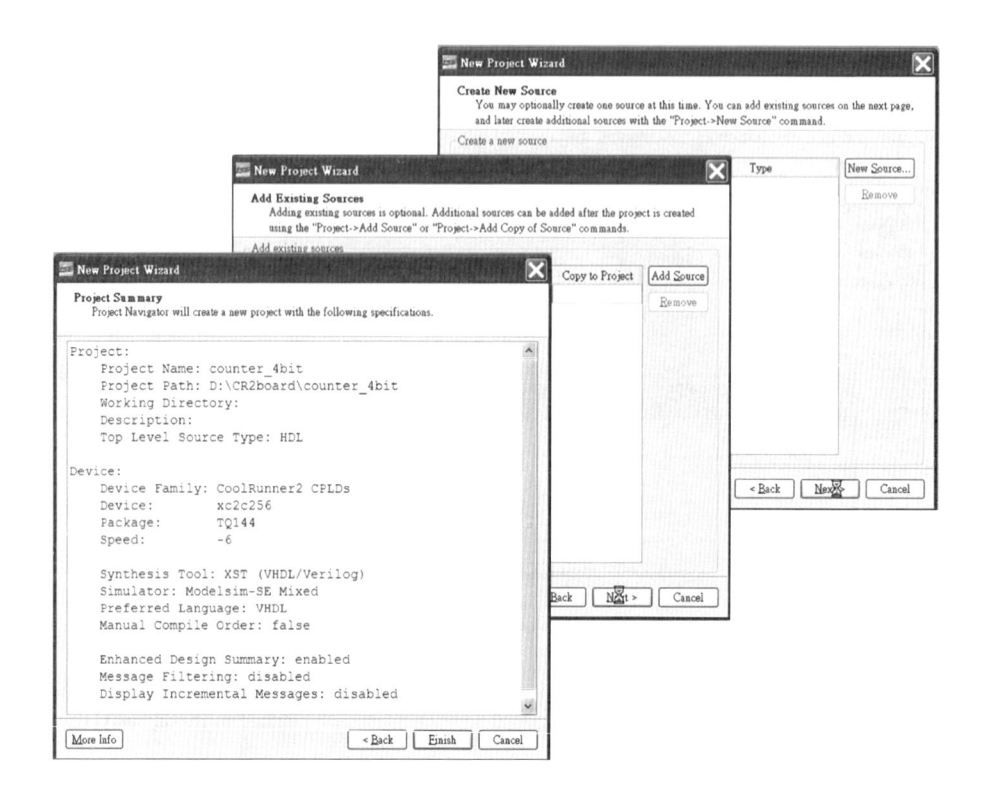

圖 G-4　　New Project Wizard 處理後的 Project Summary

2. 第二個階段：是決定 counter_4bit 的內容，從 New Source Wizard 開始決定檔案是 VHDL 和它的 I/O 名稱，及 BIT 還是 BIT_VECTOR？如圖 G-5 所示。

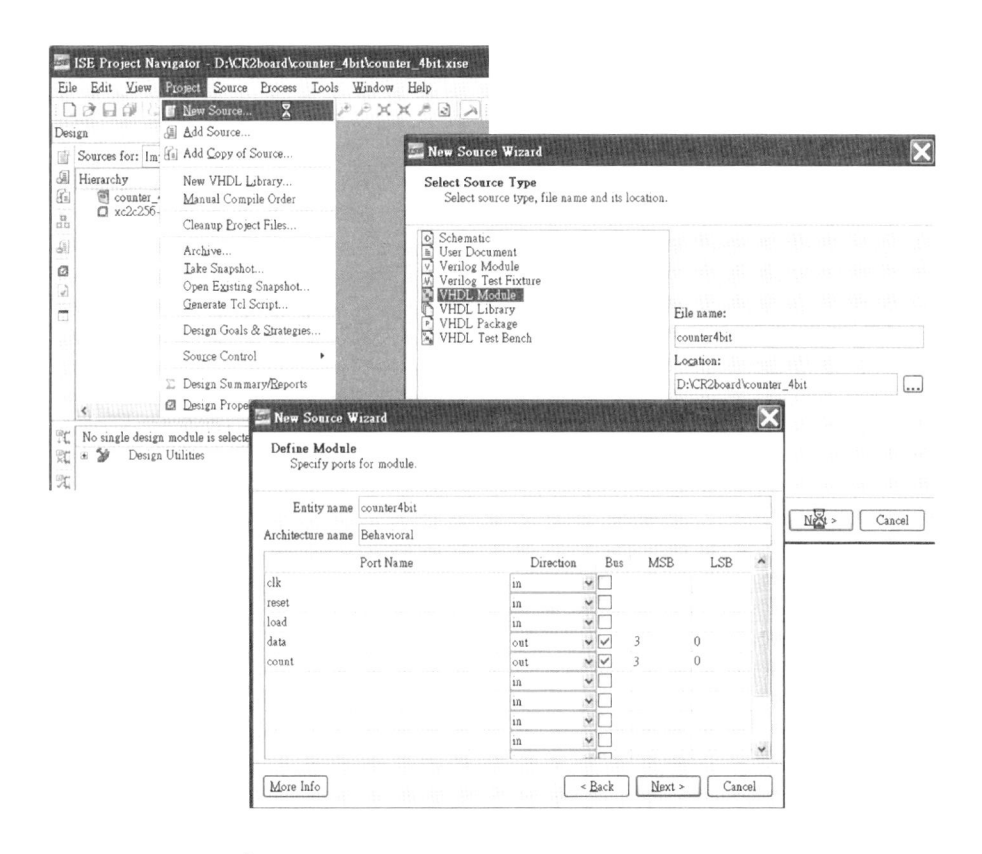

圖 G-5　　Counter_4bit.vhd 檔和它的 I/O

New Source Wizard 的 Summary 為 counter_4bit 的 I/O 清單，如圖 G-6 所示。

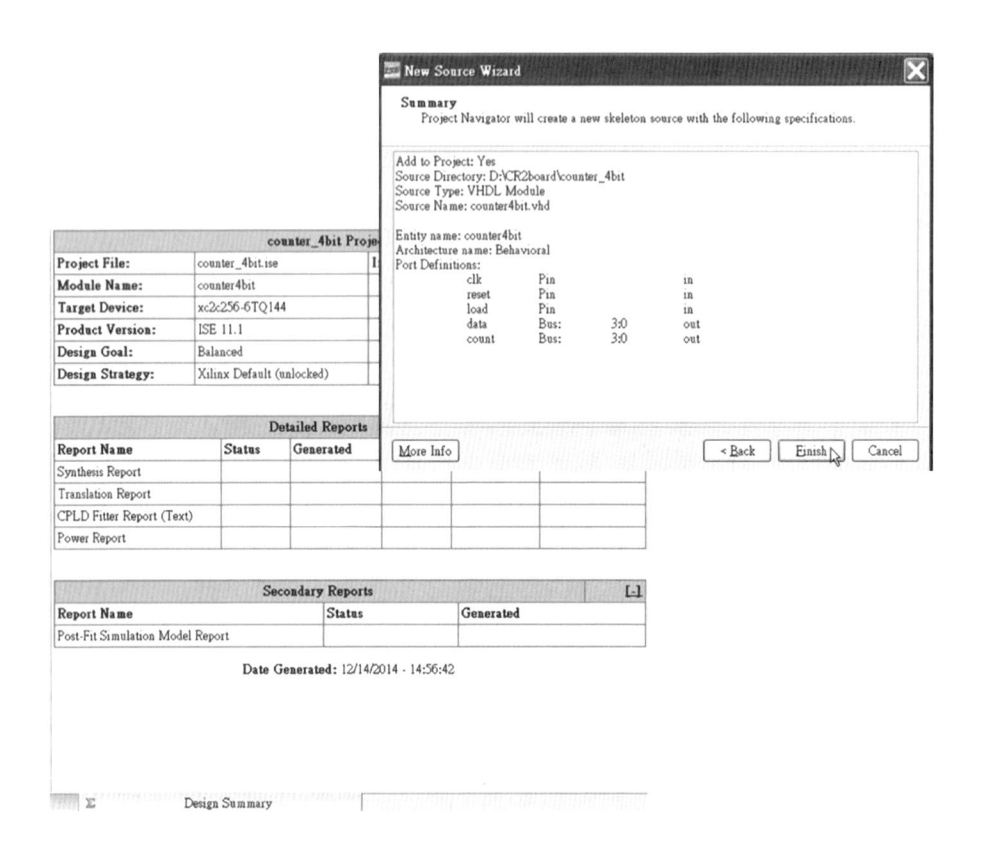

圖 G-6　counter_4bit 的 I/O 清單

ISE Project Navigator 依據圖 G-6 的 I/O 清單提供圖 G-7 檔案上半部 Line 19~43。其中 Line 41~42 Architecture 部份是空白。圖 G-7 檔案下半部 Line 41~52 是這個 counter_4bit 的設計結構，由讀者填寫上。

```
19 ------------------------------------------------------------
20 library IEEE;
21 use IEEE.STD_LOGIC_1164.ALL;
22 use IEEE.STD_LOGIC_ARITH.ALL;
23 use IEEE.STD_LOGIC_UNSIGNED.ALL;
24
25 ---- Uncomment the following library declaration if instantiating
26 ---- any Xilinx primitives in this code.
27 --library UNISIM;
28 --use UNISIM.VComponents.all;
29
30 entity counter4bit is
31     Port ( clk : in  STD_LOGIC;
32            reset : in  STD_LOGIC;
33            load : in  STD_LOGIC;
34            data : out STD_LOGIC_VECTOR (3
35            count : out STD_LOGIC_VECTOR (
36 end counter4bit;
37
38 architecture Behavioral of counter4bit is
39
40 begin
41
42
43 end Behavioral;
44
45
```

```
30 entity counter4bit is
31     Port ( clk : in  STD_LOGIC;
32            reset : in  STD_LOGIC;
33            load : in  STD_LOGIC;
34            data : out STD_LOGIC_VECTOR (3 downto 0);
35            count : out STD_LOGIC_VECTOR (3 downto 0));
36 end counter4bit;
37
38 architecture Behavioral of counter4bit is
39
40 begin
41     p0: process (clk, reset, load) is
42         variable cnt : STD_LOGIC_VECTOR(3 downto 0);
43     begin
44         if reset = '1' then
45             cnt := (others => '0');
46         elsif load = '1' then
47             cnt := Data;
48         elsif rising_edge(clk) then
49             cnt := cnt + 1;
50         end if;
51         count <= std_logic_vector(cnt);
52     end process p0;
53
54 end Behavioral;
```

圖 G-7　檔案的 Architecture 內容由讀者填寫上

3. 第三個階段：是 ISE Project Navigator 的 Process 下來做 Implement Design，也就是檢查是否有錯？如圖 G-8 所示。

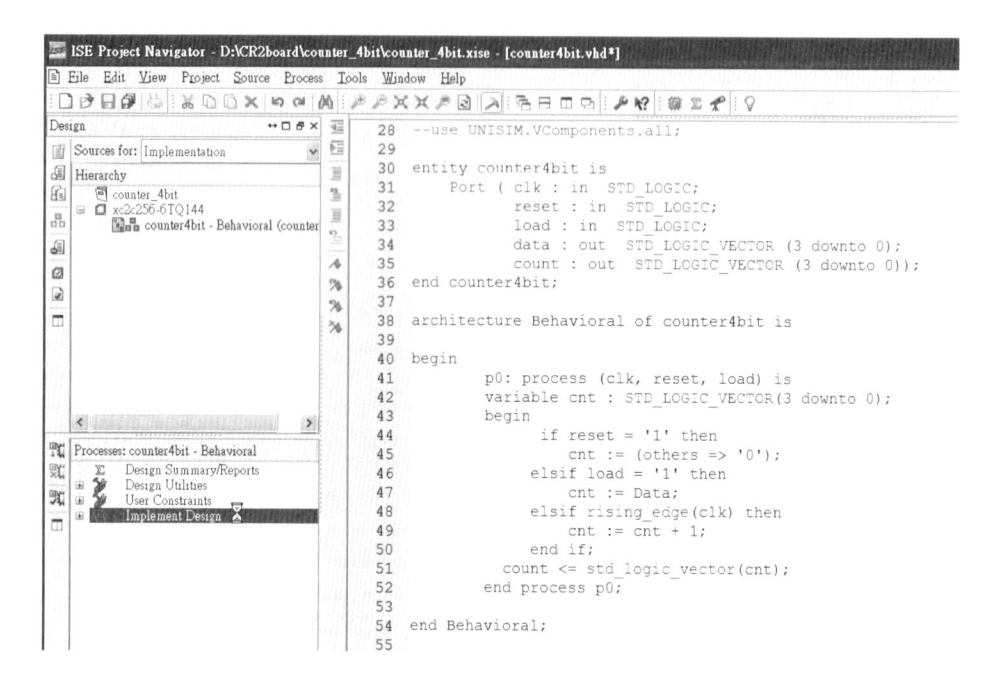

```
28    --use UNISIM.VComponents.all;
29
30    entity counter4bit is
31        Port ( clk : in  STD_LOGIC;
32               reset : in  STD_LOGIC;
33               load : in  STD_LOGIC;
34               data : out  STD_LOGIC_VECTOR (3 downto 0);
35               count : out  STD_LOGIC_VECTOR (3 downto 0));
36    end counter4bit;
37
38    architecture Behavioral of counter4bit is
39
40    begin
41        p0: process (clk, reset, load) is
42            variable cnt : STD_LOGIC_VECTOR(3 downto 0);
43            begin
44                if reset = '1' then
45                    cnt := (others => '0');
46                elsif load = '1' then
47                    cnt := Data;
48                elsif rising_edge(clk) then
49                    cnt := cnt + 1;
50                end if;
51            count <= std_logic_vector(cnt);
52            end process p0;
53
54    end Behavioral;
55
```

圖 G-8　counter_4bit 在 Process 下來做查錯

Implement Design 下有 Translate、Fit 和 Generate Program 三項被打勾，代表檢查的結果沒有錯。如圖 G-9 所示。

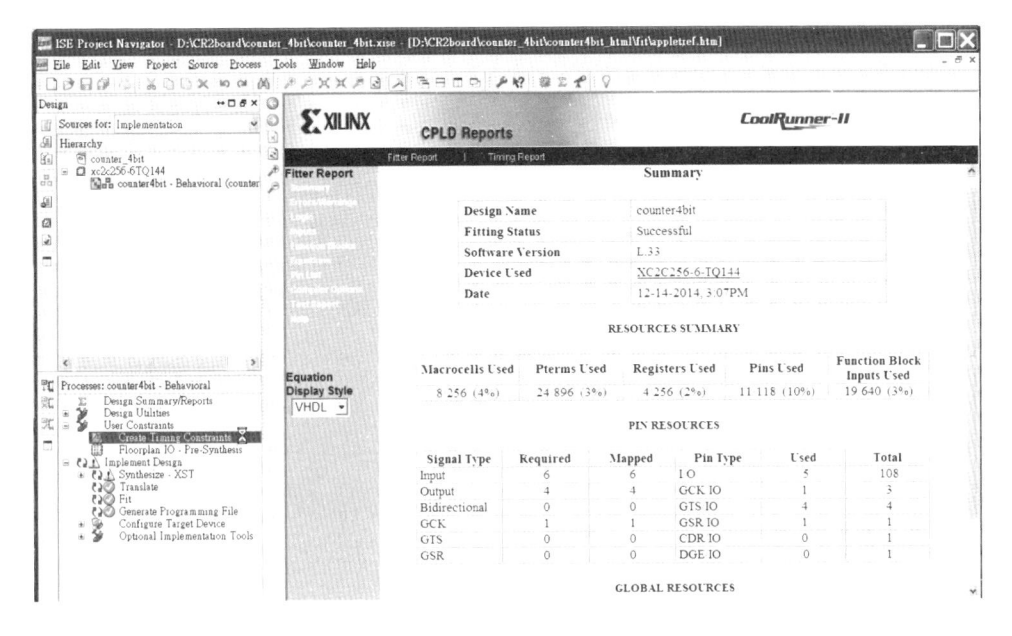

圖 G-9　counter_4bit 通過檢查

4. 第四個階段：是決定 IO 到 CPLD 的接腳安排。在 Process 下的雙擊 Floorplan IO 如圖 G-10 所示。接下來就有圖 G-11：Xilinx PACE 視窗的出現。

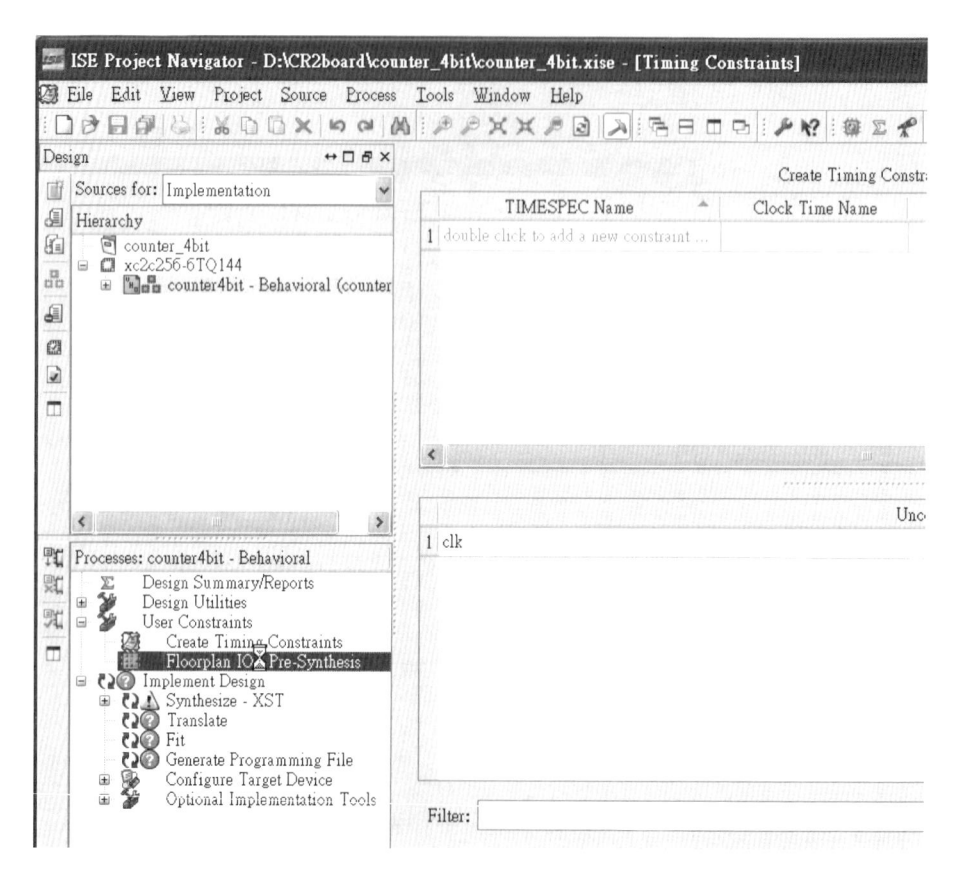

圖 G-10　　決定 counter_4bit IO 到 CPLD 的接腳安排

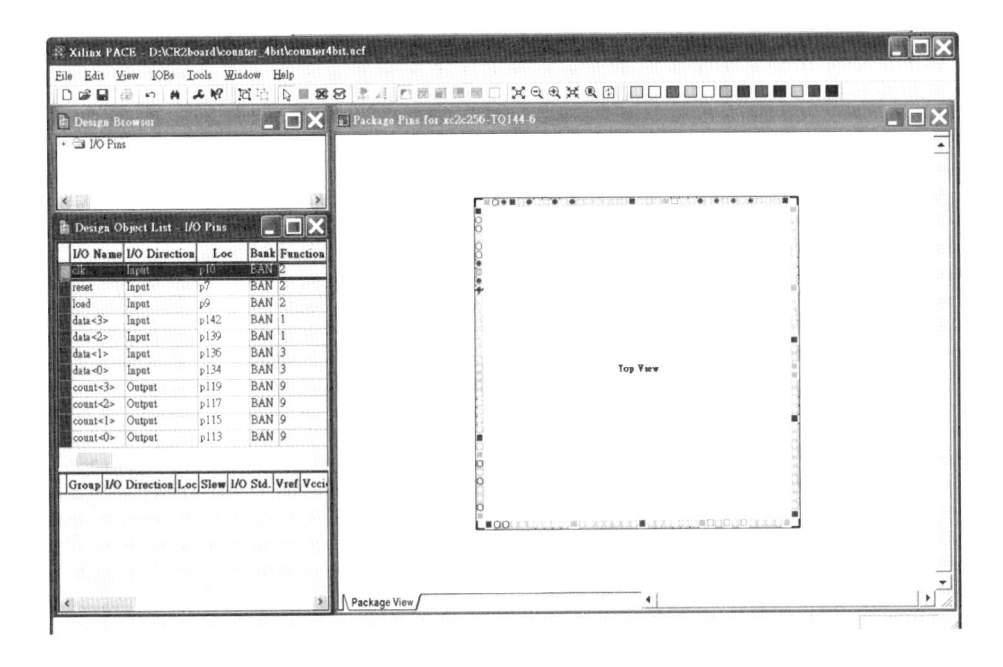

請參考圖 G-11 將 Design Object List – I/O Pins 的表格填好。

圖 G-11　Xilinx PACE 視窗與 UCF 檔的製作

圖 G-12 為 Counter_4bit 與 CoolRunner-II Starter Kit 間 I/O Pins 的關係圖，其中 J1、J2 和 J3 是 Starter Kit 的擴張接頭。用來連接到 Analog Discovery 的 Digital I/O 來做 Logic 和 Timing Analysis 之用。

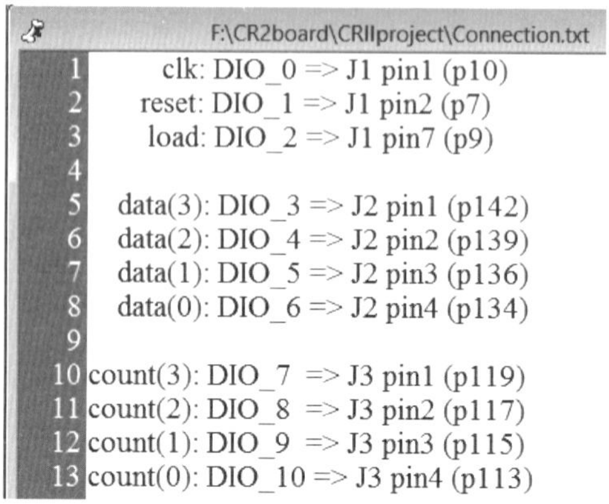

```
                    F:\CR2board\CRIIproject\Connection.txt
1        clk: DIO_0 => J1 pin1 (p10)
2      reset: DIO_1 => J1 pin2 (p7)
3       load: DIO_2 => J1 pin7 (p9)
4
5     data(3): DIO_3 => J2 pin1 (p142)
6     data(2): DIO_4 => J2 pin2 (p139)
7     data(1): DIO_5 => J2 pin3 (p136)
8     data(0): DIO_6 => J2 pin4 (p134)
9
10   count(3): DIO_7  => J3 pin1 (p119)
11   count(2): DIO_8  => J3 pin2 (p117)
12   count(1): DIO_9  => J3 pin3 (p115)
13   count(0): DIO_10 => J3 pin4 (p113)
```

圖 G-12　Counter_4bit 與 CoolRunner-II starter kit 間 I/O Pins 的關係

Save 圖 G-11 Xilinx PACE 視窗與 UCF 檔。然後回到 ISE Project 的 Implement Design，如圖 G-13。雙擊 **Implement Design** 後再一次查錯。

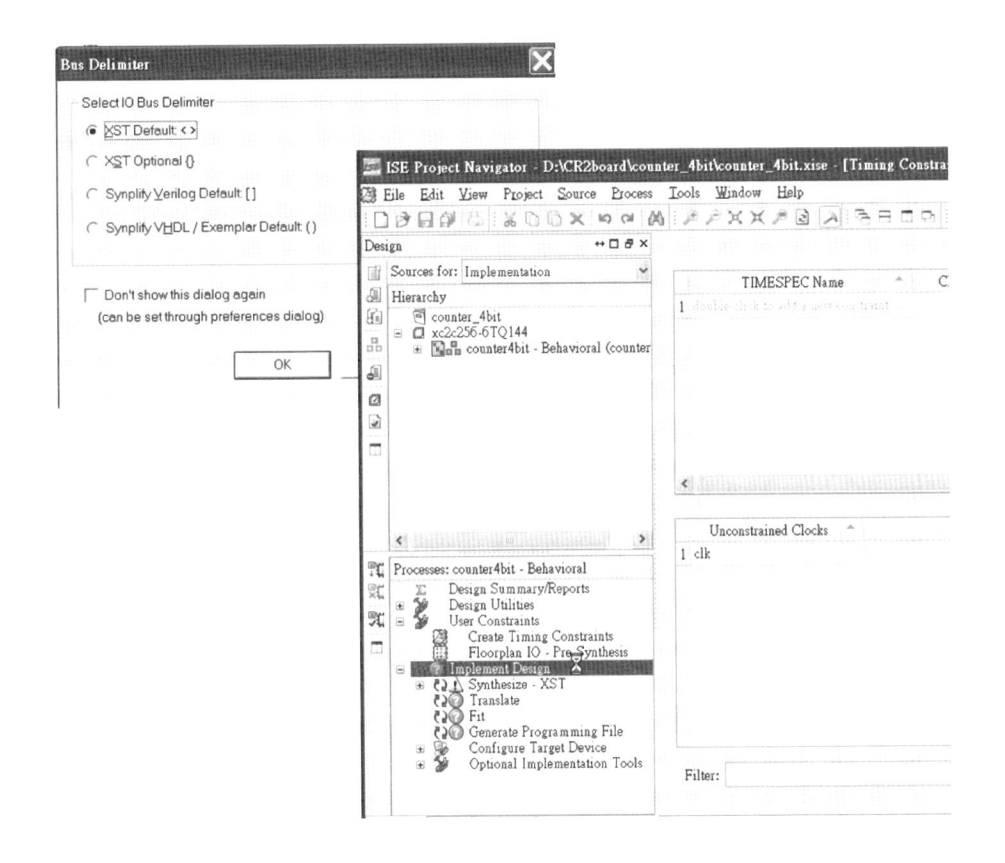

圖 G-13： Project 加入 UCF 檔後再一次查錯

242　iLAB Digital 數位電路設計、模擬測試與硬體除錯

　　這次查錯的結果，如圖 G-14 所示，OK 的話，就可以獲得 ISE Program 的結果 Counter_4bit.jed 檔。該檔是在 Counter_4bit 的檔案夾中，如圖 G-14 右方所示。

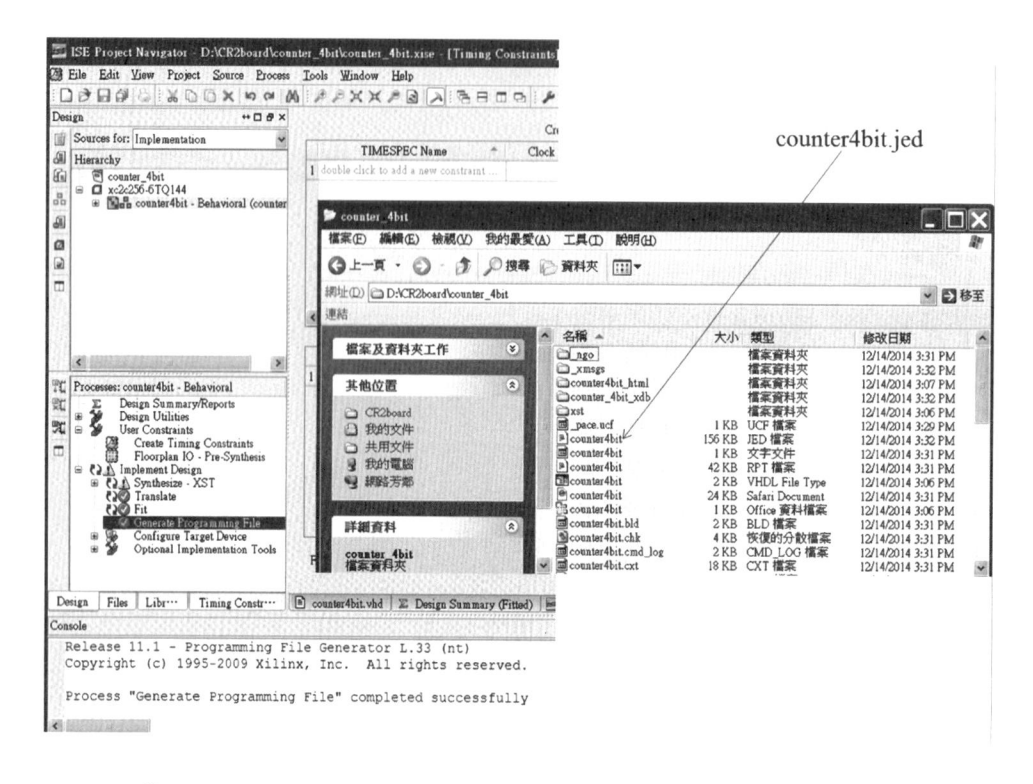

圖 G-14　ISE Program 成功的結果，獲得 Counter_4bit.jed 檔

把 Counter_4bit.jed 檔放進圖 G-2 的 CoolRunner-II Utility Window 內，單擊 Program，如圖 G-15 所示。如果圖的下端顯示 Programming Complete，就表示已成功地置入 CXC2C256 CPLD。

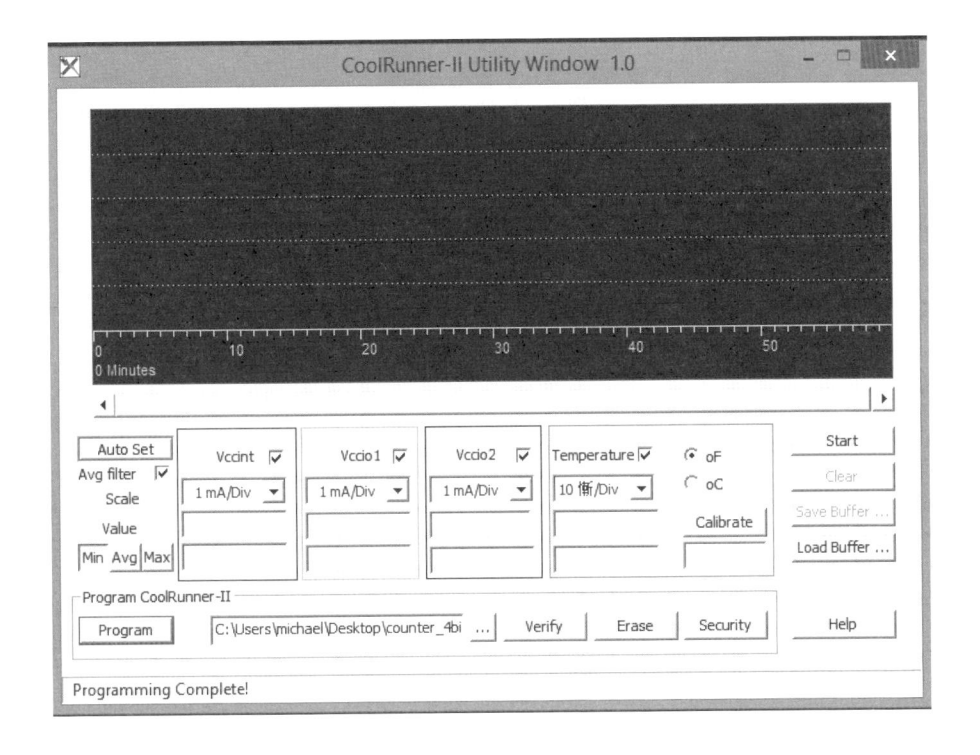

圖 G-15　CXC2C256 CPLD Program 成功

244　iLAB Digital 數位電路設計、模擬測試與硬體除錯

　　Program 成功的 CoolRunner-II Starter Board，可以依照圖 G-12 CoolRunner-II starter kit 與 Analog Discovery 間的 I/O Pins 關係，連接到 Analog Discovery 的 Logic I/O 上，如圖 G-16 所示。

圖 G-16　用 Analog Discovery 來測試 CoolRunner-II Starter Board 的設計

附錄 G　Xilinx CPLD 的電路合成　245

　　Counter_4bit 是純 Logic 電路，所以在開啓 Digilent Waveform 1 之後，如圖 G-17 所示。只須使用 Digital 部份的 Pattern/out 來做電路的 Digital Pattern Generator 作為輸入。和 Analyzer/in 也就是 Logic Analyzer 來做電路的輸出分析。

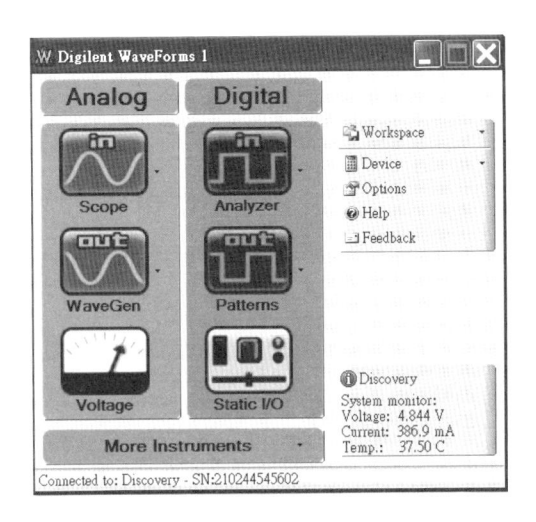

圖 G-17　開啓 Digilent Waveform 1 和選擇

　　圖 G-18 為 Digital Pattern Generator 設定，Counter_4bit 的所有輸入如 clk、reset、load 和 Bus 0 即 data (3 downto 0) 都應當包括在內。Trigger 必須選用 Analyzer 以便二者同步。

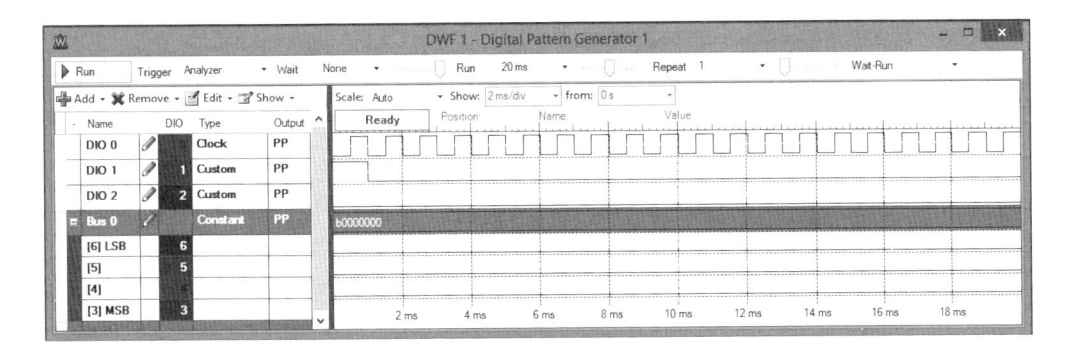

圖 G-18　Digital Pattern Generator 的設定

246　iLAB Digital 數位電路設計、模擬測試與硬體除錯

　　圖 G-19 為 Logic Analyzer 的設定，Counter_4bit 的所有輸出如 count (3 downto 0) 必須包括在內之外，還可以加入部份或全部的輸入。圖 G-19 加入了 CLK 和 reset，以便觀測訊號間的 Timing 關係。

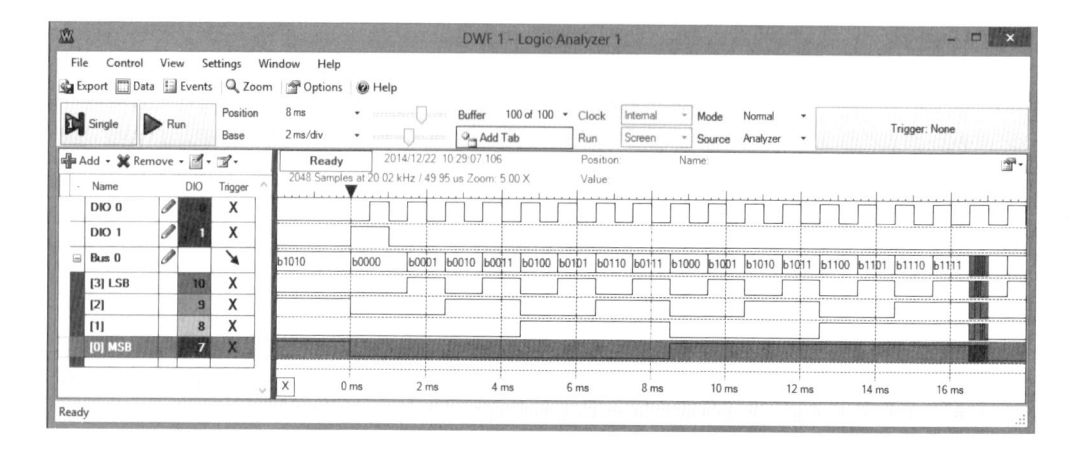

圖 G-19　測試 Counter_4bit 的輸出和部份輸入訊號的 Timing 關係

索引

中英對照

SAR　Successive Approximation Register　81

五劃

半步　Half-step　106

六劃

行為(型)　Behavior　33, 46, 56

八劃

並發邏輯　Concurrent Logic　1

九劃

計時　Timer　55
計數器　Counter　55, 69, 98

十劃

時鐘; 時序脈波　Clock　1, 93
時序邏輯　Sequential Logic　1, 21
脈衝寬度調變　Pulse Width Modulation, PWM　118

十一劃

單相　Single-phase　105

十二劃

結論　Summary　133

十四劃

裝載　Load　49

十五劃

數字模式產生器　Digital Pattern Generator　5, 15, 26, 39, 51, 62, 110
數位信號分析儀　Digital Signal Analyzer　5, 15, 26, 39, 51, 62, 110
編輯　Compilation　139

十八劃

鎖相環　Phase Locked Loop　93
雙相　Two-phase　106

二十劃

麵包板　Breadboard　4, 14, 25, 38, 51, 61, 99

英中對照

B

Behavior 行為（型） 33, 46, 56
Breadboard 麵包板 4, 14, 25, 38, 51, 61, 99

C

Clock 時鐘；時序脈波 1, 93
Compilation 編輯 139
Concurrent Logic 並發邏輯 1
Counter 計數器 55, 69, 98

D

Digital Pattern Generator 數字模式產生器 5, 15, 26, 39, 51, 62, 110
Digital Signal Analyzer 數位信號分析儀 5, 15, 26, 39, 51, 62, 110

H

Half-step 半步 106

L

Load 裝載 49

P

Phase Locked Loop 鎖相環 93
Pulse Width Modulation, PWM 脈衝寬度調變 118

S

Sequential Logic 時序邏輯 1, 21
Single-phase 單相 105
Successive Approximation Register SAR 81
Summary 結論 133

T

Timer 計時 55
Two-phase 雙相 106

248